# Oxygen Dynamics in the Chesapeake Bay

## *A Synthesis of Recent Research*

Edited by

David E. Smith
*University of Virginia and*
*Virginia Sea Grant College*

Merrill Leffler
Gail Mackiernan
*Maryland Sea Grant College*

*Produced by Maryland and Virginia*
*Sea Grant College Programs*

A Maryland Sea Grant Book
College Park, Maryland

Published by the Maryland Sea Grant College in cooperation with the Virginia Sea Grant College.

Publication of this book is supported by grant NA86AA-D-SG042 from the National Oceanic and Atmospheric Administration to Maryland Sea Grant and Virginia Sea Grant.


Printed in the United States of America.



Book design by Sandy Harpe

Maryland Sea Grant College
Publication UM-SG-TS-92-01
Virginia Sea Grant College
Publication VSG-92-01
Library of Congress Card Catalog Number: 91-068257
ISBN 0-943676-50-9

For information on Maryland Sea Grant books, contact:

Maryland Sea Grant College
University of Maryland
College Park, Maryland 20742

*Printed on recycled paper*

# Contents

# Acknowledgments

While a great deal of the information presented in this book resulted from studies sponsored by the National Oceanic and Atmospheric Administration, the authors have incorporated data from work supported by other federal agencies and state governments, in particular, the Environmental Protection Agency, the National Science Foundation, the State of Maryland and the Commonwealth of Virginia.

The editors wish to thank the authors of the individual chapters for their contributions, as well as the critical reviewers who provided valuable comments on draft versions of these chapters. We also want to acknowledge and thank the scientists in Maryland and Virginia who contributed to this volume through discussions and the sharing of unpublished information with the authors.

Support for the Hypoxia/Anoxia Program and for this publication were provided through NOAA's National Sea Grant College Program, grant number NA86AA-D-SG-006 to the University of Maryland Sea Grant Program and grant number NA86AA-D-SG-042 to the Virginia Sea Grant College Program.

# Hypoxia Studies in the Chesapeake Bay: An Overview

The Chesapeake Bay is the nation's largest estuary, its most complex and historically its most productive. The Bay's main stem ranges two hundred miles, from the mouth of the Susquehanna River in Maryland to the ocean in Virginia, covering some 4,400 square miles. Increasingly, Chesapeake Bay has been subjected to the impacts of accelerated population growth, from massive volumes of sediment eroding from the 64,000 square mile watershed, human and industrial wastes pumped into the Bay's rivers and streams, and thousands of tons of nutrients annually coming from agricultural lands, sewage treatment plants and acid rain. The consequences have been degraded water quality, massive loss of sea grasses and severely diminished stocks of commercial fish and shellfish.

To restore the integrity of the Bay's ecosystem, resource managers must have a sophisticated understanding of the biological, chemical and physical processes that regulate the Bay's health. A major goal of ongoing scientific research in the Chesapeake is to advance that understanding so resource management agencies can better develop informed action plans.

A recurring condition in Chesapeake Bay — and an important measure of the estuary's health — has been hypoxia, a depletion of oxygen in bottom waters that begins in spring and lasts through summer. There is some controversial evidence that the extent and severity of hypoxia (and anoxia, the complete lack of dissolved oxygen) have been increasing in recent years, possibly as a result of human-induced changes in the Bay ecosystem. While scientists have long known that this phenomenon results from a number of factors — natural circulation patterns, excessive nutrients, microbial action in sediments — they lacked sufficient understanding of how the interactions among biological, chemical and physical processes affected the Bay's dissolved oxygen levels. Because they could not detail the relative contributions of natural and human effects that contribute to hypoxia, it has been difficult to develop protective management guidelines for minimizing the impact from human activities. Many management decisions have relied more on informed guesswork and generalized models.

In 1985, the Maryland and Virginia Sea Grant programs, with support from the National Oceanic and Atmospheric Administration, undertook a long-term program of integrated research aimed at detailing how biochemical and physical processes between sediments and overlying waters regulate hypoxia.

Over the last six years, this cooperative effort has advanced our understanding of the dynamics of the Bay ecosystem considerably. A number of the scientific and management findings summarized here are playing important roles in defining research needs and assisting resource management agencies.

## Major Scientific Findings

*1. Hypoxic and anoxic conditions generally commence in late spring and extend through the summer.* Critical to initiation of these conditions is the annual spring inflow of fresh water down tributary rivers. The result of snow melt and heavy spring rains, this spring runoff increases stratification, lowers Bay salinity, affects the residence time of water and materials, and delivers nutrients and sediments to the Bay.

*2. The role of nutrient loading in causing hypoxia in Chesapeake Bay should be further examined in light of an apparently significant relationship found between the volume of hypoxic water and spring river flow.* On one hand, this relationship might indicate that the reported 1950-1980 increase in oxygen depletion resulted from a natural cycle in river flow (via stratification suppressing reaeration of the bottom waters), and that increased nutrient loading has not yet altered the balance. While the rate of water-column oxygen consumption appears relatively invariant from year to year, the rate of reaeration of lower-layer waters is strongly dependent upon seasonal stratification set up by spring runoff. On the other hand, the expectation remains that biochemically mediated processes control an underlying trend of oxygen depletion. Careful assessment of the relative roles of physical and biological factors will be required to resolve this uncertainty and reveal the key controls. Care is required because many of the important effects of freshwater inflow — nutrient delivery, stratification and gravitational circulation — are highly coupled.

*3. The amount of dissolved oxygen in lower layers of Chesapeake Bay is strongly influenced by stratification.* Stratification, in turn, is controlled by the balance between fresh water runoff (creating stratification) and mixing (acting to destroy stratification). Estuarine stratification responds to fresh water runoff on three primary time scales. A freshet from the Susquehanna drives a seaward surge of flow in the middle reaches of the Bay within one week; this surge will increase stratification by decreasing the salinity in the upper layer. The same freshet will produce an increase in the two-layer estuarine circulation and an associated increase in stratification in the mid-Bay approximately one month later. On a seasonal time scale, spring runoff builds up a "buoyancy reservoir" of fresh water in the Bay. This reservoir is then available throughout summer for rapid restratification after wind-mixing events. More importantly, the buoyancy reservoir provides a mechanism for the spring-runoff influence on stratification to extend the three-month interval from peak runoff to the height of the summer anoxia season.

*4. Hypoxia occurs when the rates of oxygen consumption in the lower layers of Chesapeake Bay are greater than the resupply from vertical and longitudinal transport.* Resupply occurs through vertical mixing from the upper layers and horizontal movement of deep oxygenated water from the lower reaches of the Bay. River runoff increases stratification and, in turn, decreases vertical exchange, but it also creates a counteracting reaeration through increased gravitational circulation. Stratification is the dominant process, and this advective reaeration is secondary. However, circulation supplying aerated deep water from the southern Bay creates an oxygen minimum in the vertical and confines the horizontal minimum to the northern portion of the deep trough of the Bay.

*5. Spatial and temporal patterns of nitrogen, phosphorus and silicon distribution and recycling rates suggest that on an annual basis phytoplankton production within the entire Bay is nitrogen-limited, while on smaller spatial and temporal scales phosphorus and/or silicon are also probably limiting.* Nutrient species of nitrogen and silicon are removed more efficiently than phosphorus from the water column Calculations indicate that external nitrogen sources are capable of supporting only 10 to 20% of the phytoplankton production on a yearly basis. The remainder, therefore, must be the result of recycled nutrients within the Bay itself.

*6. A seasonal shift occurs in the source of nutrients to the euphotic zone: during spring, high nutrient loading occurs because of river runoff; during summer, it occurs as a result of recycling processes in sediments.* As a result of these processes, nitrate inputs from all sources peak in spring, while phosphate and ammonium inputs peak in the summer.

*7. Peaks in phytoplankton biomass and productivity are out of phase: chlorophyll concentrations in the water column are highest in spring, while areal primary productivity reaches its maximum in summer.* In spring, the phytoplankton assemblage is characterized by high biomass and relatively slow turnover, whereas in summer it is characterized by low biomass and rapid turnover. Grazing of phytoplankton by herbivores during summer helps regulate phytoplankton biomass. This seasonal pattern coincides with a shift in size composition, favoring larger algae (diatoms) in spring and smaller phytoplankton forms in summer months. For example, picophytoplankton primary production increases from 2% of the total primary production in winter to greater than 20% after the spring bloom.

*8. Oxygen depletion is linked to nutrient loading through the accumulation, deposition and decomposition of phytoplankton biomass.* Synoptic observations of environmental factors from 1984 to 1988 in the mesohaline reach of the Bay indicate that oxygen depletion is not directly related to the role of phytoplankton productivity *per se,* but it is related to net production (photosynthesis minus respiration) of the plankton community and associated accumulation and deposition of organic matter.

*9. The spring peak in phytoplankton biomass is the result of new nutrient inputs, and the consumption of this organic matter, primarily by bacteria and protozoa, serves to initiate seasonal oxygen depletion in Chesapeake Bay.* The rate of oxygen depletion in spring is a function of temperature and availability of labile organic carbon and shows small interannual variability.

*10. The accumulation of phytoplankton biomass in spring and the low grazing pressure by herbivores such as copepods at this time results in the funneling of available organic carbon to bacteria.* Bacteria in the Bay have been shown to increase their turnover rate of organic substrates as substrate concentrations increase. Consequently, the bacteria are poised to take quick advantage of this increase in available organic nutrients. As this occurs, the traditionally defined trophic flow from phytoplankton-to-zooplankton-to-fish (and macroinvertebrates) is disrupted and instead carbon flows from phytoplankton-to-bacteria-to-protozoa. This decoupling or redirection of energy flow can have significant impacts on the entire ecosystem.

*11. Bacterial abundances in Chesapeake Bay are exceptionally high during the spring and summer, at times approaching 20 million cells ml$^{-1}$.* During the late spring and early summer, rates of bacterial production peak, approaching 1250 mg carbon m$^{-2}$d$^{-1}$. Biomass calculations indicate bacterial production is often 30%, and can exceed 60%, of the phytoplankton biomass during late summer. Bacterial abundances and metabolic rates suggest that particulate organic matter is insufficient to sustain populations and that dissolved organic matter such as dissolved monosaccharides serve as a primary source of energy for bacterial growth. Future investigations into the chemical nature, concentrations and availability of dissolved organics as energy sources need to be undertaken.

*12. Bacteria and organic matter play a significant role in the consumption of oxygen in Chesapeake Bay as corroborated by the fact that respiration of small particles (<1 to 3 μm) accounted for 60 to 100% of the total plankton metabolism.* Vertical zones of bacterial abundance, bacterial metabolism and specific organic compounds occur in the water column. Mid-water concentration and metabolic maxima are common, with peaks often observed near the pycnocline. Such distributions result in a localization of oxygen consumption, and the presence of a "bacterial plate" acting as a biological barrier against oxygen flux into bottom waters.

*13. Biogeochemical cycling of sulfur is important in the demand on dissolved oxygen: while production of sulfides occurs primarily in the sediments, if anoxic conditions already exist in deep waters, sulfide diffusion into the water column results in oxygen consumption near the pycnocline.* During the summer, measurements of sulfide oxidation in this region indicate a mean oxygen consumption rate of 9 mg L$^{-1}$d$^{-1}$ . This rate is comparable to measured aerobic heterotrophic demand beneath the pycnocline under aerobic conditions.

*14. Deposition of particulate organic matter from the euphotic zone to the Bay bottom provides most of the food which supports*

***metabolism of benthic organisms and plankton and which leads, in conjunction with stratified conditions, to hypoxia in bottom waters.*** The magnitude, food quality and timing of sedimenting particulate organic matter is controlled by variations in nutrient supply rates, organic production and consumption processes of the plankton community.

*15. **Most of the nutrients recycled in summer are derived from deposition of particulate organic matter fixed in spring.*** Rates of nutrient (nitrogen, phosphorus, silicon) recycling from metabolic processes in the benthos are sufficient to support 20 to 80% of the nutrients required for phytoplankton growth.

*16. **The natural processes of nitrification and denitrification which remove nitrogen from the benthos are inhibited in Chesapeake Bay under hypoxic conditions; the resulting increased recycling of nitrogen supports additional algal production of organic matter and bacterial oxygen consumption.*** Denitrification coupled to nitrification in Bay sediments is responsible for removing 25 to 30% of nitrogen entering the estuary from the watershed and atmosphere; this rate is low compared with other estuaries in which denitrification accounts for 50 to 55% removal of nitrogen.

*17. **Compared to other coastal ecosystems, the relative biomass of benthic macrofauna in the Bay appears to be lower than expected from observed levels of phytoplankton abundance and production.*** This anomaly, which may also result in depressed demersal fisheries, is most likely related to loss of habitat associated with seasonal hypoxia in large regions of the Bay.

*18. **Under present conditions, intense grazing by suspension-feeding benthic macrofauna (mostly bivalves) in certain shallow well-mixed areas of the Bay appears to result in relatively reduced abundance of phytoplankton.*** This situation is particularly evident in low salinity regions of the main Bay and its tributaries. It is possible that this effect was more widespread in former times when oyster populations, which were vastly more abundant throughout the upper estuary, may have controlled phytoplankton growth. Other factors, however, may mitigate this suspension-feeding effect. First, greater numbers of benthic macrofauna would lead to increased defecation and excretion that could accelerate nutrient recycling, thereby enhancing production; second, bivalve grazing on zooplankton could reduce grazing pressure on phytoplankton, which could also lead to greater production.

*19. Interannual changes in benthic community metabolism and nutrient recycling processes are directly related to variation in rates of nutrient loading from the watershed.* There is also indirect evidence of a small carry-over of organic matter and associated nutrients from one year to the next. Bay sediments do not, however, appear to be vast storage areas for labile organic matter, such that reductions in nutrient loading to the estuary should result in significant decreases in rates of benthic oxygen consumption and nutrient recycling within one to two years.

## Management Implications of Scientific Findings

*1. Chesapeake Bay is particularly susceptible to the hypoxic and anoxic effects of nutrient enrichment because of its natural, two-layer gravitational circulation.* This two-layer circulation which characterizes the Bay much of the time creates conditions which (1) favor the retention of nutrients in particulate form, allowing this organic matter to be efficiently used and reused before being buried or exported to the sea, and (2) effectively reduces the vertical mixing of oxygenated surface waters with oxygen-depleted bottom waters.

*2. Recent observations of a significant relationship between summer volume of hypoxic water and spring river flow suggests that previously reported historical increases in anoxia associated with eutrophication should be viewed with caution.* Detection of such an increase would require careful separation of the impact on dissolved oxygen levels by natural physical processes and biological and chemical processes. Unfortunately, this separation does not involve the simple extraction of linear component influences from time-series records. The underlying physics as well as the dependence of the biological and chemical processes on physical forcing has been shown to be richly non-linear. Given that a major goal of the Chesapeake Bay clean-up program is reduction of nutrients from land-based sources which are considered to be the primary cause of increased hypoxia, it is important to determine unambiguously whether or not there is even a trend. Such a goal is essential for evaluating the success of nutrient reduction efforts.

*3.* ***Recent evidence indicates a residual effect of nutrient inputs to the Bay which operates on time scales of one month to two years.*** For example, summer rates of sediment nutrient recycling are highly correlated with algal deposition to sediments during the previous months of spring. Further, it appears that relatively high nutrient inputs during a high river-flow year support relatively high rates of primary production and benthic oxygen consumption through the following year. The actual mechanisms controlling these temporal lags between coupled processes, however, are not well understood. The question of how long and at what magnitude this residual effect can influence Bay ecology is central to predicting the impact of reducing nutrient loading to the estuary as well as predicting the outcome of management strategies for improving the Bay's environmental condition.

*4.* ***Currently, the mesohaline Bay region appears to be nutrient-saturated in that seasonal cycles of oxygen depletion in bottom water appear to vary little with changes in nutrient loading under current conditions of climate and watershed development.*** Consequently, until nutrient loads and corresponding phytoplankton biomass are reduced below some unknown level or threshold, where dissolved oxygen demand exceeds rates of dissolved oxygen replenishment, hypoxia and anoxia will continue to occur.

*5. Restoration of the Bay's lost denitrification potential would reduce nitrogen levels by rejuvenating this important natural removal mechanism.* Current hypoxic conditions in the summer inhibit the natural biochemical processes of nitrification and denitrification that convert nutrient forms of inorganic nitrogen salts to biologically unavailable gaseous forms of nitrogen; this is because nitrification cannot occur without oxygen. This effect constitutes a positive feedback loop that allows inorganic nitrogen salts to be recycled so as to enhance algal production in the Bay. The higher algal production leads to further oxygen depletion from bottom water. In reverse, however, this positive feedback can accelerate the process of reducing hypoxia due to the restoration of lost denitrification potential, which has not been realized because of hypoxic inhibition of nitrification and denitrification processes.

*6. The effects of decoupling the phytoplankton-to-zooplankton-to-fish and macroinvertebrate flow of energy to a phytoplankton-to-bacteria-to-protozoa flow may have reduced natural production of commercially important species.* For example, the shunting of primary production (phytoplankton) to bacteria may reduce the energy available to benthic filter feeders such as oysters because of large energy losses associated with microbial metabolism. Management actions which reduce hypoxic conditions in the Bay will probably result in improved production of benthic macrofauna and certain demersal fish populations, both through provision of habitat and reestablishment of a diverse, benthic-feeding community of fish and invertebrates.

*7. Decomposition rates for sediment organic matter and associated recyclable nutrients and oxygen demand are sufficient to preclude significant long-term accumulation over decades.* Thus, nutrient concentrations, algal biomass and biological respiration rates should decrease rapidly within several years following reductions of nutrient inputs to the Bay.

*8. Numerical models are important for simulating and forecasting the effects of nutrient loading on anoxia.* The models must include (1) accurate representation of the balance between river runoff and saltwater intrusion (buoyancy fluxes) acting to create stratification; (2) detailed representation of planktonic trophic interactions which control sedimentation of particulate organic matter; (2) detailed description of planktonic trophic interactions; (3) benthic nutrient recycling processes which are, themselves, strongly affected by oxygen conditions; (4) benthic macrofauna communities which affect rates of nutrient recycling and particulate organic matter removal from the water column.

***9. Experience in recent years has clearly indicated the value of generating long-term data sets over appropriate time scales; such data sets are necessary for developing an understanding of the forces which regulate the Bay ecosystem and affect the linkages between system components.*** While current monitoring programs provide extensive spatial coverage, a complementary program of more measurements at a limited number of sites (i.e., to increase temporal coverage) would help further define relationships between benthic and pelagic system components and aid better quantification of circulation processes. Paleostratigraphic techniques hold promise for retrospective analysis and are generating provocative assessments of the relative impact of climate and nutrients on hypoxia.[1]

***10. Continuous measurements of higher frequency physical and biological processes are attainable by using enhanced technology such as remote sensing and in-situ moorings designed to record selected environmental conditions.*** This kind of high-frequency coverage would vastly improve our ability to determine the proper spatial and temporal scales for detailed study.

---

[1] Cooper, S.R. and G.S. Brush. 1991. Long-term history of Chesapeake Bay anoxia. Science.

# Introduction

The Chesapeake Bay has traditionally served as a source of livelihood for thousands of commercial watermen and seafood processors and as a commercial artery linking Europe to two of the largest ports on the East Coast — Hampton Roads, Virginia and Baltimore, Maryland. Increasingly, the Bay attracts new home owners, businesses, large numbers of recreational boaters and sports fishermen.

In addition to natural changes that result from storms and changing sea level, the Chesapeake ecosystem, like all heavily populated coastal systems, has been subjected to the impacts of an expanding population, industrial development, modern agriculture, waste disposal, and numerous other pressures. Many of these changes manifest themselves in a deterioration in water quality and a decline in the abundances of once-rich natural resources.

With population growth in the Chesapeake region continuing — by the year 2020 twice as many people are expected to live in the watershed as in 1950 — prospects for improving water quality and for protecting Bay waters are daunting. To provide help for the critical management efforts ahead, the National Oceanic and Atmospheric Administration initiated a research program in 1985, coordinated through the Sea Grant Programs in Virginia and Maryland, to investigate the mechanisms responsible for what appeared to be increasing episodes of hypoxia (the near absence of dissolved oxygen) in the bottom waters of the Bay. Hypoxia and anoxia (the absence of all oxygen) are both indicators of the Bay's health and phenomena which are detrimental to its living resources, especially those which make up the vital benthic environment, namely bottom-dwelling fish, oysters, clams and crabs.

While a number of state and federal monitoring programs and research projects continue to provide new understanding about the Bay ecosystem, the purpose of this interdisciplinary research effort was to support a program that would investigate, over a period of several years, the biological, chemical and physical interactions that contribute to the apparent spread of hypoxia and anoxia in the Bay, and to determine its effects on the estuary's productivity.

The depletion of oxygen in the deep waters of Chesapeake Bay has been documented since the early 1900s (Sale and Skinner 1917; Newcomb and Horne 1938). With the spring thaw in the watershed comes a major inflow of fresh water (called a freshet) down tributaries to the Bay. Buoyant and light, these waters flow over denser incoming ocean water. In this two-layer flow a boundary or stratified layer called the pycnocline forms. The degree of stratification depends on the intensity of the freshet, which, in turn, depends on natural climatic events — for example, the amount of snowfall in winter, the extent of spring rains and the intensity of winds. Runoff rate and periodicity is generally considered to have increased approximately 30% since clearing of the Bay's forest cover began (Bosch and Hewlett 1982).

With the spring flows and runoff from the land come associated sediments, organic material and nutrients. As the Bay begins to warm and daylight is longer, phytoplankton grow more abundantly. In spring, growing faster than they can be consumed by slower-growing zooplankton grazers, phytoplankton biomass increases.

Eventually, phytoplankton die and sink through the water, decay and are oxidized. In deeper waters, below the pycnocline, this oxygen demand results in an oxygen deficit: bottom waters become hypoxic and may eventually experience anoxia. This phenomenon begins in the Bay in spring and lasts generally until mid fall.

It has long been understood that oxygen depletion is related to several factors: (1) the production of large amounts of organic matter, (2) the heterotrophic metabolism of that organic matter and (3) a deficit in the rate of deep water reaeration. Nevertheless, scientists have lacked an understanding of the specific sources and nature of the organic matter; the nature and activities of the organisms which actually consume dissolved oxygen and organic matter; the role of chemical oxidation in the water column and at the sediment/water interface; the relative importance of physical circulation; and the spatial and temporal distribution of these forcing parameters. Furthermore, there has been little detailed understanding of the potential effects of hypoxia on the food web, except a general understanding that there would be reduced habitat for populations of important biological and commercial resources.

This interdisciplinary research program brought together coastal chemists, biologists and physicists from Virginia and Maryland to define these important processes. Appendix A provides a list of research projects supported between 1985 and 1990. While data have been taken throughout the Bay, from Annapolis, Maryland to the Great Wicomico River in Virginia, much of the research has focused on the mesohaline region of the Bay — that portion where the salinity is generally between 5 and 18 ppt. Measurements were taken at stations along an east-west, cross-Bay transect (Figure 1). Called the Chop-Pax transect because it extends from the Choptank River to the Patuxent River, these locations were chosen for several reasons: bottom water anoxia was common, the Bay is narrow enough to allow a research vessel to sample transect stations within a reasonably synoptic time period, and two marine laboratories are located nearby, which minimizes the time necessary to reach the sampling transect.

Like all sampling programs, the design did have limitations. For example, in Chesapeake Bay the spring bloom of phytoplankton typically occurs in the surface layer, reaching peak biomass concentrations at points southward of the Chop-Pax transect. However as the phytoplankton populations decline, age and sink into the bottom waters, residual landward currents common in two-layer estuarine systems transport the phytoplankton biomass up the Bay into the mesohaline region where maximum oxygen consumption occurs (Figure 2). Therefore, in an attempt to maximize data collection in the vicinity of known anoxia development, researchers have had to

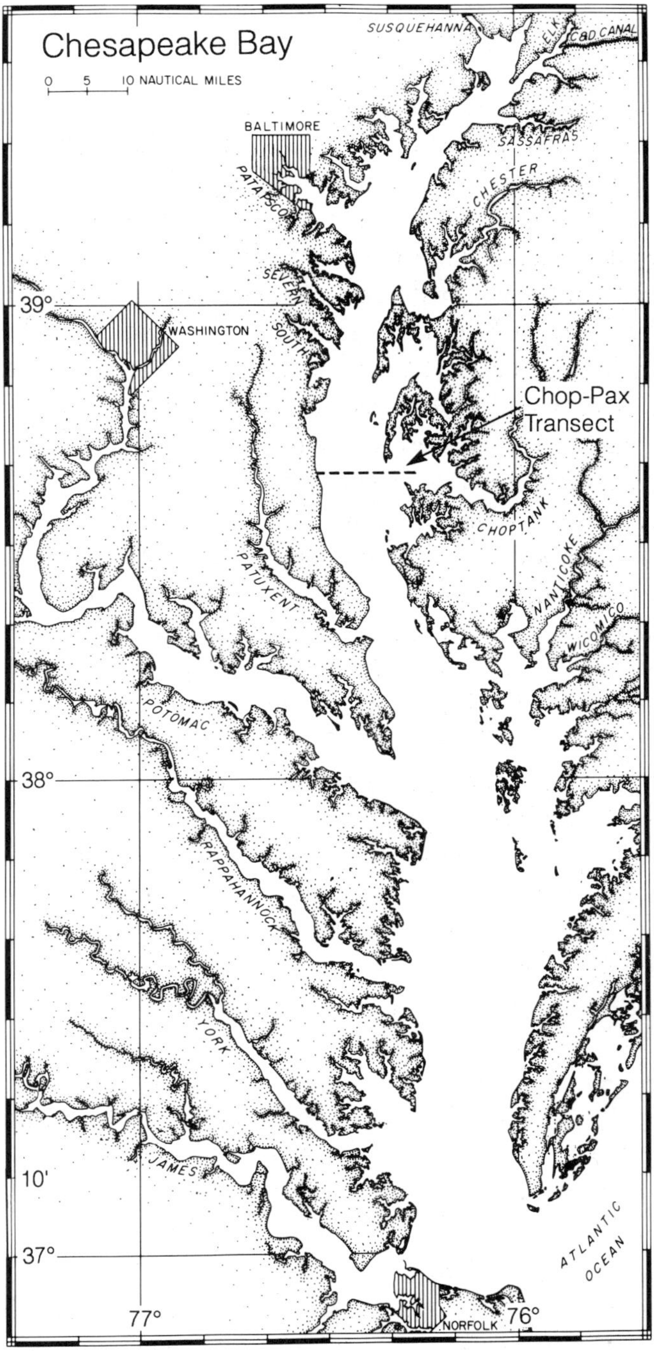


*Figure 1. Chop-Pax transect in the mesohaline region of Chesapeake Bay.*

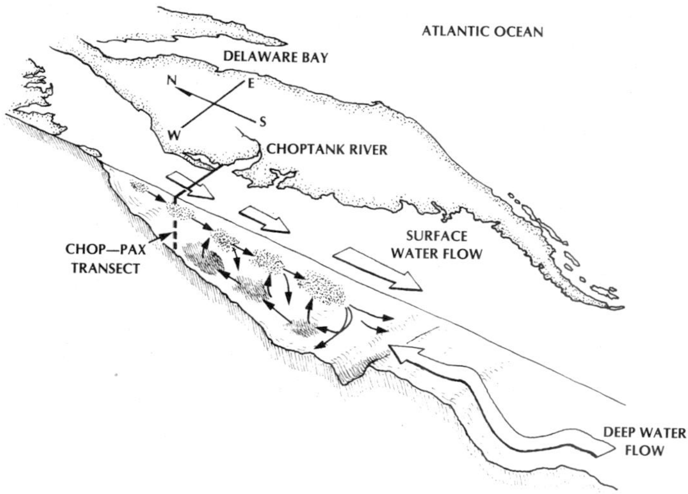


*Figure 2. Cross-sectional view of the Chesapeake Bay's two-layer estuarine flow.*

compromise by not fully sampling the water column throughout the area where peak phytoplankton biomass develops.

Nevertheless, the data generated from this transect created regular vertical "snapshots" of what was occurring within this region of the Bay. These snapshots — when coupled with data collected both north and south of the transect (though on a less consistent basis), and the EPA and state-sponsored monitoring data — have served to explain the creation and maintenance of low dissolved oxygen in bottom waters of Chesapeake Bay. The four chapters of *Oxygen Dynamics in the Chesapeake Bay* address these issues, each from a different perspective. The authors highlight major trends and the gradually developing consensus on the processes occurring in the Bay. As a summary, this volume should be of interest to the Bay's scientific community as well as to resource managers.

William Boicourt examines the physical aspects of Chesapeake Bay in order to distinguish the role and relative importance of the Bay's circulation in the creation of the low dissolved oxygen layer. Thomas Malone then focuses on the role of water column processes, including the importance and interaction of nutrients, phytoplankton and zooplankton. The significance of microbial populations and the processes involving them is the theme of Robert Jonas's contribution. Finally, Michael Kemp and Walter Boynton discuss the role and importance of benthic-pelagic coupling on Chesapeake Bay dissolved oxygen.

As each author's acknowledgments and references indicate, this book represents the cumulative efforts of many scientists in Maryland and Virginia. While it summarizes research activity to date, it raises important questions that should help guide continuing research. The entire research project itself represents the kind of successful activity that is possible when federal and state agencies and universities work together cooperatively to address environmental problems.

## References

Bosch, J.M. and Hewlett. 1982. A review of catchment experiments to determine the effect of vegetation changes on water yield and evapotranspiration. J. Hydrol. 55:3-23.

Newcombe, C.L. and W.A. Horne. 1938. Oxygen-poor waters of the Chesapeake Bay. Science 88:80-81.

Sale, J.W. and W.W. Skinner. 1917. The vertical distribution of dissolved oxygen and the precipitation by saltwater in certain tidal areas. J. Franklin Inst. 184(Dec):837-848.

# Influences of Circulation Processes on Dissolved Oxygen in the Chesapeake Bay

*William C. Boicourt*
Horn Point Environmental Laboratory
Center for Environmental and Estuarine Studies
University of Maryland System, Cambridge, Maryland 21613

## Introduction

The pattern and motion of Chesapeake Bay waters exert strong influences on the distribution of dissolved oxygen within this large and complex estuary. Some aspects of this influence are direct and obvious, such as the development of a barrier that restricts the supply of oxygen to the bottom waters during spring and summer. This barrier, or pycnocline, results from stratification set up by light, fresh water from rivers flowing seaward over dense, salty water flowing into the estuary from the adjacent ocean. Stratification limits vertical exchange, and can control oxygen depletion in the bottom waters by suppressing reaeration from the well-oxygenated overlying waters.

Other aspects of circulation-process control of oxygen depletion appear far more indirect and perplexing. For example, once new nutrients are delivered to the estuary via the spring runoff, the transport and dispersion of these nutrients, their uptake in the spring phytoplankton bloom, and their ultimate fate during decomposition of this organic matter (requiring oxygen) are biochemical pathways mediated at many stages by the motion of Bay waters. Other examples of such coupled and complex physical-biological processes are productivity bursts driven by wind-forced upwelling of nutrient-rich waters, biological responses to shorter term destratification and reaeration events, and alteration of predator-prey relationships by motion and stratification.

While our ability to describe and model complex circulation processes in the Bay is evolving at a rapid pace, we still are at a loss to answer some seemingly simple questions about trends in dissolved oxygen. In 1983, the Environmental Protection Agency concluded that the summertime oxygen depletion in the Bay had increased in both geographical extent and duration, as a result of human activities. The primary basis for this judgment was the record from a series of cruises carried out by researchers at the Chesapeake Bay Institute from 1950 to 1980. Recent attempts to link oxygen-depletion volumes indicated by this record to their root causes have led some investigators to seriously question the conclusion of a spatially and temporally growing hypoxia in the Chesapeake Bay. Seliger and Boggs (1988), for example, have found a correlation between the volume of oxygen-depleted water in summer and the amount of spring runoff from the Susquehanna River. Such a correlation might indicate that the reported 1950 to 1980 increase in oxygen depletion arose from a natural cycle in river flow, and that human activities have not yet altered the balance. The correlation does underscore the inadequacy of our current understanding of the relationship between nutrient input and oxygen depletion in the Bay. Perhaps even more frustrating is the lack of an unequivocal demonstration of a trend in hypoxia. In light of the extraordinary management efforts proposed to reduce the nutrient loading on the Bay ecosystem, at least this level of understanding is necessary if we are to evaluate the impact of human activity (including these reductions) on dissolved oxygen.

The primary determinant of the amount of oxygen-depleted water in the summer Bay appears to be stratification. The spring surge in river runoff produces a light, low-salinity blanket of water that overlies the heavier, higher-salinity water entering from the ocean. This blanket impedes reaeration of the (oxygen-starved) lower-layer water by suppressing vertical exchange. The degree of suppression depends on the density difference between these two layers. While this insulation of the lower-layer water is a physical, circulation process, oxygen depletion is fundamentally a biochemical process (Taft et al., 1980). How, then, can it be both? As will be discussed later, a likely explanation for this paradox is that the processes of oxygen consumption may be relatively invariant from year to year while the rate of vertical mixing varies significantly with spring runoff and the resulting stratification in the estuary.

An assessment of the biological influence on oxygen depletion requires separating the physical component of variability in order to detect lower-level biochemical signals. However, this separation requires that physical processes be described to an accuracy that is at the limit of present observational and analytical techniques. Accuracy is necessary because we would like to separate human influence from natural cycles, and because the dominant biological and physical processes are highly coupled. River inflow, for instance, delivers nutrients to the Bay and fuels oxygen depletion (Malone, this volume). At the same time, freshwater inflow controls stratification and circulation in the estuary. Deciphering the causal connections in this coupled system will be difficult, though essential, if we are to establish trends unambiguously and provide the scientific support for Bay restoration efforts.

The following discussion about the role of physical processes in hypoxia begins with a brief overview of our current understanding and then moves to questions of major interest: does freshwater inflow control oxygen depletion through stratification or nutrient delivery? What is the role of horizontal advection in the Bay and in the tributaries? What are the vertical exchange processes? What can the physics of the system explain about biological and chemical rates: can we distinguish between water column and sediment demand for oxygen? What level of understanding is necessary to determine the effects of management efforts to resolve the oxygen depletion efforts? What are the priority research efforts to attain this understanding?

## Chesapeake Bay Circulation

Studies of physical processes in Chesapeake Bay have played a pivotal role in the evolution of our present understanding of estuarine circulation dynamics (Beardsley and Boicourt 1981). The first models were developed in the 1950s, when researchers from the Chesapeake Bay Institute were trying to answer questions about transport mechanisms of oyster larvae. With innovative current measurement techniques, they conducted an extensive series of cruises on the James River. Sampling was sufficiently intensive that it could produce stable averages in the presence of tidal oscillation. From these measurements, Pritchard (1952, 1953, 1954) developed a description of the two-layer estuarine, or gravitational circulation, with fresher upper-layer water moving seaward over high-salinity water moving toward the head of the estuary (Figure 1). Rattray and Hansen (1962) developed an analytical model of estuarine circula-

*Figure 1. Two-layered circulation pattern with zone of maximum turbidity.*

tion based on Pritchard's formulation of the dynamic and salt-balance equations. Agreement between the analytical solutions and the James River observations enhanced the influence of both studies. In the years following, researchers have expanded on Pritchard's classical description by examining other dynamical effects, particularly the role of wind, mixing and bottom friction.

Although early investigators showed a keen awareness that strong wind forcing affects estuarine circulation, wind-driven motion was seldom incorporated into models. Instead, wind effects were often invoked to explain model discrepancies. The development of quantitative descriptions of wind-driven currents requires current-meter deployments of sufficient duration to allow the passage of a series of atmospheric low-pressure systems. From these data, wind-driven motions could be statistically separated from the slowly varying gravitational circulation.

Weisburg and Sturges (1976) employed such long-term current measurements to examine the effects of wind transport in Narragansett Bay, a partially mixed estuary. They showed that over short time intervals (2 to 8 days), wind-driven motions can easily dominate the longitudinal transport of water. Weisburg (1976) also showed that a proper separation of the wind-driven motion from gravitational circulation is difficult with short records.

Farmer (1976) addressed the role of wind in fjord estuaries and accurately modeled Farmer and Osborn's (1976) observations of fluctuations in upper-layer thickness in Alberni Inlet. Soon after, Elliott (1978) conducted a year-long current meter mooring in the middle reaches of the Potomac River estuary. Estuarine circulation, he discovered, was affected both by wind forcing acting locally over the estuary and by wind forcing acting at a distance. For a tributary such as the Potomac River, this remote forcing arises from wind-forced changes in sea level in Chesapeake Bay proper. Water moves in or out of the Potomac to match the tributary's level with that of the main stem. Local and remote wind forcing can override the two-layer gravitational circulation for intervals up to a week. The resulting combination can be a one-, two- or three-layer flow. In Elliott's year-long records from the Potomac, the classical two-layer flow structure was present only 43% of the time. Wang and Elliott (1978) found that local circulations in Chesapeake Bay tributaries could be influenced by remote forcings well beyond the Bay, on the continental shelf. In turn, sea level at the entrance of Chesapeake Bay could be affected by continental shelf waves remotely generated by storms at distances on the order of 500 km from the Bay entrance.

Within Chesapeake Bay's shallow water tributaries, most of the remotely forced circulation is set up by the oscillatory fluctuations of the Bay's sea level which are driven by wind stress and by gradients in atmospheric pressure. The primary mode for these motions is a rise and fall of the upper Bay water level called the quarter-wave seiche. In this mode, the water level near the mouth of the Bay remains nearly constant (a node), and the sea surface slopes upward or downward toward the head of the Bay (an antinode). These oscillations, with a period of approximately 2 days (Wang and Elliott 1978; Wang 1979), can be seen in low-pass filtered (to remove the semi-diurnal and diurnal tides) water level records (Figure 2), with maximum amplitude at the head of Chesapeake Bay and decreasing amplitude toward the node near the mouth of the Bay.

The magnitude of the currents created by the seiche (Figure 3) is at a maximum, not at the nodal point, which is inside the Bay (Chuang and Boicourt 1989) but nearby at the entrance to Chesapeake Bay, where there is a constriction in cross-sectional area. These oscillations are driven by strong winds and atmospheric pressure fluctuations, and may be triggered by wind stress acting over only the southern portion of the Bay. On the other hand, the highly damped character of these motions and the shallowness of the Bay indicate that they are primarily forced, rather than free, oscillations. If they are, then the integrated wind stress over the entire length of the Bay is active in producing these motions. In any case, these oscillations constitute the dominant subtidal variability in flow and sea level in the Bay. Olson's spectral model (1986) of the subtidal response of the Bay to meteorological forcing predicted a peak variance frequency in volume transport at 0.4 cycles $d^{-1}$, for all three locations examined within Chesapeake Bay. This oscillation is the quarter wave seiche response.

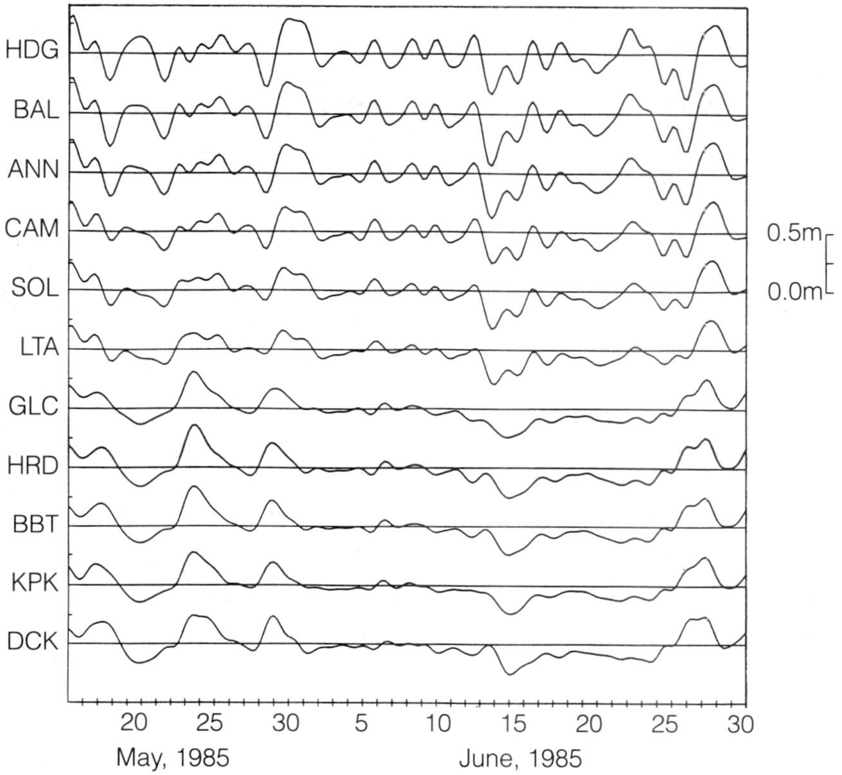


*Figure 2. Low-pass (Lanczos 34-h) filtered water level records in Chesapeake Bay and the adjacent continental shelf for May and June 1985. From top to bottom, the gauge designations are Havre de Grace, Baltimore, Annapolis, Cambridge, Solomons, Lewisetta, Gloucester Point, Hampton Roads, Bay Bridge Tunnel, Kiptopeake and Duck, North Carolina. The two-day period of the Chesapeake Bay quarter-wave seiche is evident in the records from the northern portion of the Bay, near the antinode.*

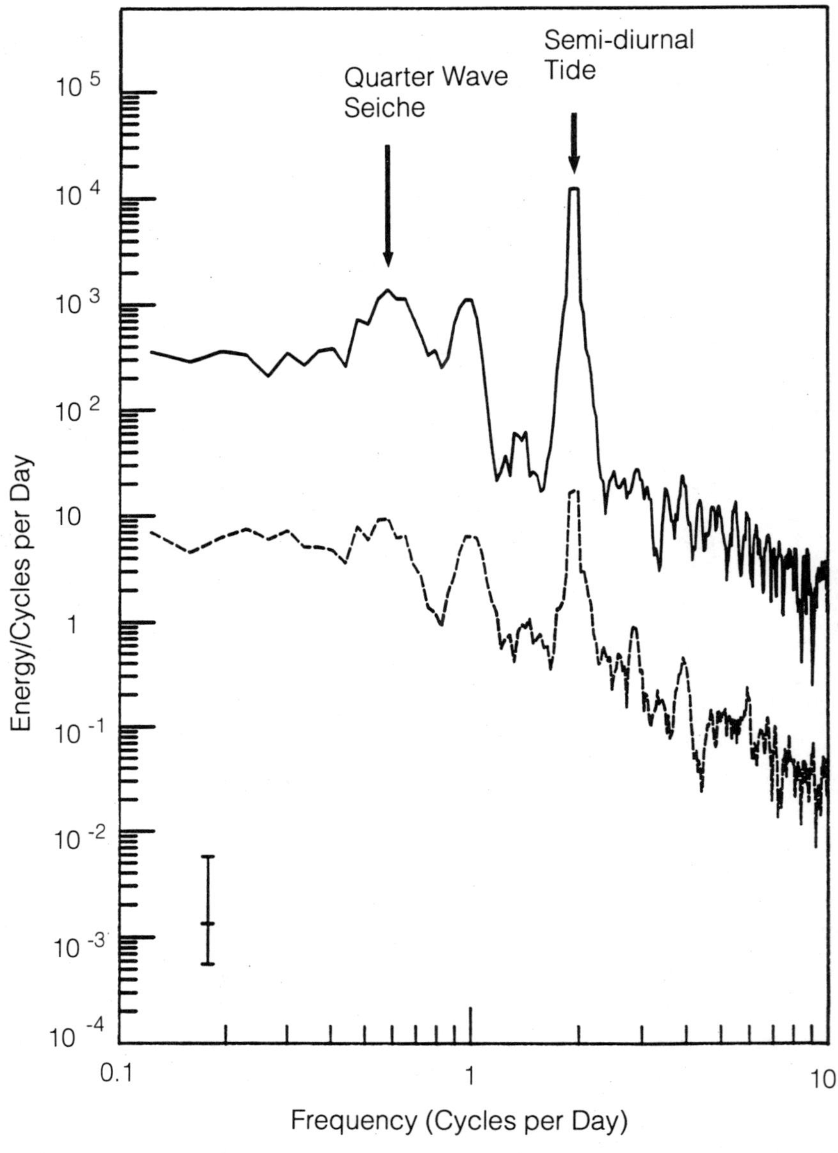


*Figure 3. Spectra of velocity (solid line) and salinity (dashed line) from an instrument at 2.4 m depth in the entrance to Chesapeake Bay during April 1986. Arrows indicate peak at quarter-wave seiche period and the semidiurnal tide.*

### *Mixing*

Estuaries, including the Chesapeake Bay, have provided key laboratories for experimentation of transport and mixing processes of the ocean. Many of these studies, which have been reviewed by Beardsley and Boicourt (1981) and Itsweire and Phillips (1987), have addressed the critical unknown in the circulation of both the estuary and the open ocean — turbulent mixing. Knowledge of mixing processes where water, salt, heat and momentum are stirred and mixed is crucial to unraveling estuarine physics, especially because mixing drives estuarine circulation. Mixing raises the potential energy of the water column by transferring salt from the lower layer to the upper layer, thereby raising the center of mass of the water column. This potential energy then provides the source for the kinetic energy of circulation.

The classical view of estuarine mixing has depended primarily on turbulence generated at the estuarine boundaries (the sea surface, the bottom and the shoals along the shoreline) by the oscillatory motions of tidal currents. In recent years, more attention has been paid to the action of wind stress in mixing estuarine waters. A problem with this boundary mixing, whether generated by tides or driven by wind, is that our understanding of the mechanisms for transferring this mixing from the boundaries to the interior is uncertain. The process of turbulent entrainment, where turbulence generated at the boundary propagates away from the boundary, mixing nonturbulent fluid as it moves, does not provide an explanation for the observed vertical distribution of temperature, salinity and density in partially mixed estuaries. While entrainment serves to sharpen the pycnocline by creating two well-mixed layers, observed profiles indicate a broad transition region between the upper mixed layer and the bottom mixed layer. This inability for entrainment models to explain the diffuse pycnocline structure has led researchers to look for internal mixing modes that can produce the observed profiles. Although many candidate processes have been offered — internal-wave breaking, Kelvin-Helmholtz instabilities, or dynamic instabilities such as suggested by Pollard et al. (1973) — most have a common mixing source: internal velocity shear. Shear is the spatial gradient of water motion, such as the upper layer of the estuarine circulation moving seaward over the landward-moving lower layer. In their study of wind-induced destratification in Chesapeake Bay, Goodrich et al. (1987) identified a critical value of the bulk Richardson number $R_i$ (representing a ratio between stratification and velocity shear), whereupon rapid vertical mixing takes place. The coincidence of this critical value with intervals of reduced stratification, or complete destratification in the Bay provides evidence for an internal mixing process that operates in conjunction with mixing generated at the boundaries.

Whatever the mechanisms through which the action of wind stress and bottom stress is transferred into internal mixing, a partitioning can be made between the relative proportion of mixing energy derived from wind stress and from tidal interaction with the bottom topography. For example, Farmer and Freeland (1983) developed an energy budget for Knight Inlet, a fjord estuary, and found that the amount of mixing energy from wind stress was about one-third of that produced by tidal currents interacting with the bottom topography. While such partitioning between the effects of wind and bottom friction has not been carried out for Chesapeake Bay, the expectations are that wind mixing is relatively more important because the Bay's tidal range is significantly lower than Knight Inlet (Goodrich et al. 1987). In the tributary estuaries in the southern portion of Chesapeake Bay, bottom-generated mixing appears to play a dominant role, such that spring tidal currents create a destratified water column (Haas 1977). Such spring-neap stratification variability is less evident in the northern Chesapeake Bay tributaries, presumably because the tidal amplitude is lower.

When the strong, oscillatory motion of tidal currents interact with features of the bottom topography, the result is not only the generation of turbulence for mixing, but also the production of flow structures such as jets and eddies. If these flows are averaged over a tidal period, then the average measured by stationary current meters (Eulerian) usually does not agree with averages measured by drifters moving with the water (Lagrangian). The difference between the Eulerian and Lagrangian means is called the Stokes drift. As Ianiello

(1977) pointed out, the magnitude of these Stokes transport velocities can amount to the same order as the Eulerian mean velocities.

Flow structures such as jets and eddies produced by the tide flowing over ridges, sills, lateral shoals, or other topographic features seldom disappear in tidal averages. The reason for their appearance in tidal averages is that these features form as narrow, high-velocity flows downstream of topographic disturbances. If such a disturbance happens to be upstream from a current-meter mooring during one phase of the tide, then the measurements will reflect these higher velocities, which are elevated above the surrounding tidal flows. Upon reversal of the tide, the jets and eddies will no longer form on the current-mooring side of the topographic feature and, therefore, will no longer be seen by the current meter as elevated velocities. These flows can produce velocities that are sufficiently elevated above the surrounding flows that they dominate the tidal average. The shears and rotating flows associated with these tidal means constitute what Zimmerman (1978) calls residual vorticity. In addition to the dynamical effects of tidal-generated vorticity, there are significant effects on the patterns and rates of dispersion.

An example of three-dimensional flow features producing Stokes drifts and residual vorticity can be seen in measurements from an intensive array of 39 current meters moored in a short (8-km) section of the Potomac River estuary (Boicourt 1982). Lateral shoals and headlands produce jets and eddies in the upper layer (Figure 4) that are decoupled from the flows in the lower layer (Figure 5). During ebb tide, the headland (Ragged Point) on the southern shore of the estuary creates a strong jet and lee eddy seaward of the promontory. Similar structures are generated by Piney Point, on the northern shore. Corresponding jets and eddies are generated upriver during flood tide. While the average of all the upper-layer flow is seaward, locally these tidally generated features can dominate the Eulerian means, to the point where the classical central channel can also influence the means, although the influence is only obvious near the bottom. One-month mean flows (Figure 5) are consistent with a lateral pycnocline tilt (associated with the rotation of the earth), such that the upriver flows below the pycnocline are stronger

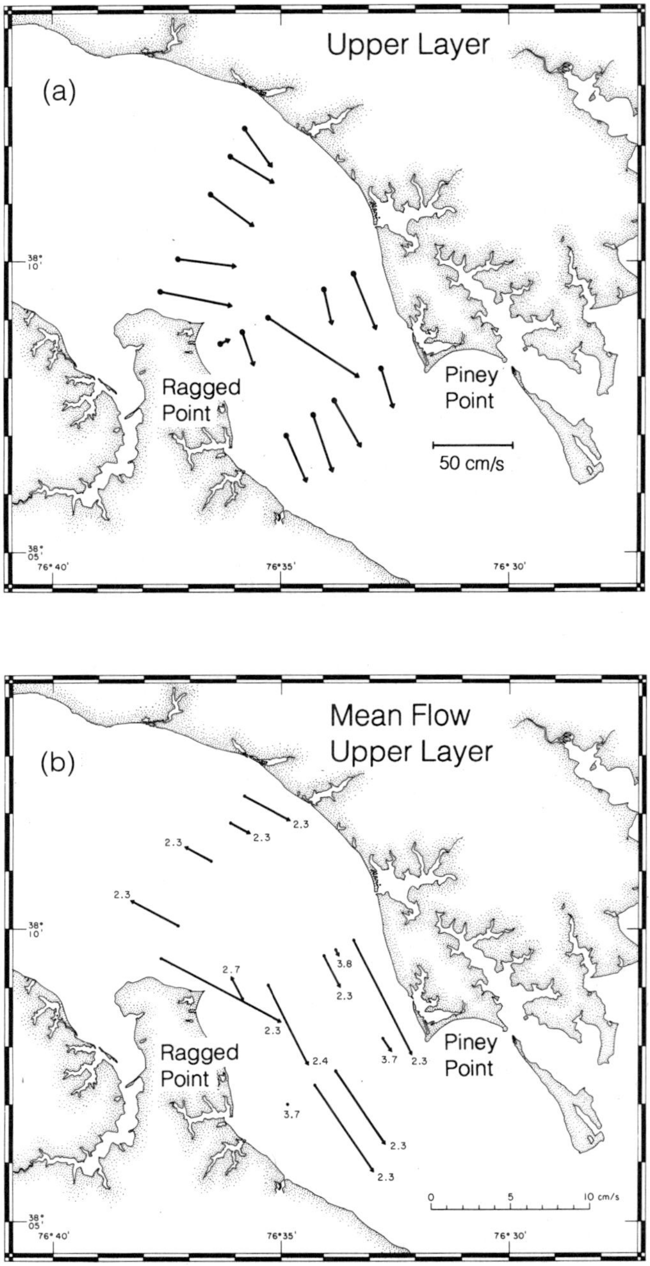


*Figure 4. (a) Instantaneous velocity measured at a depth of 2.4 m near the end of ebb tide at 1500 on May 31, 1978 in the Potomac River estuary. (b) One-month mean velocities.*

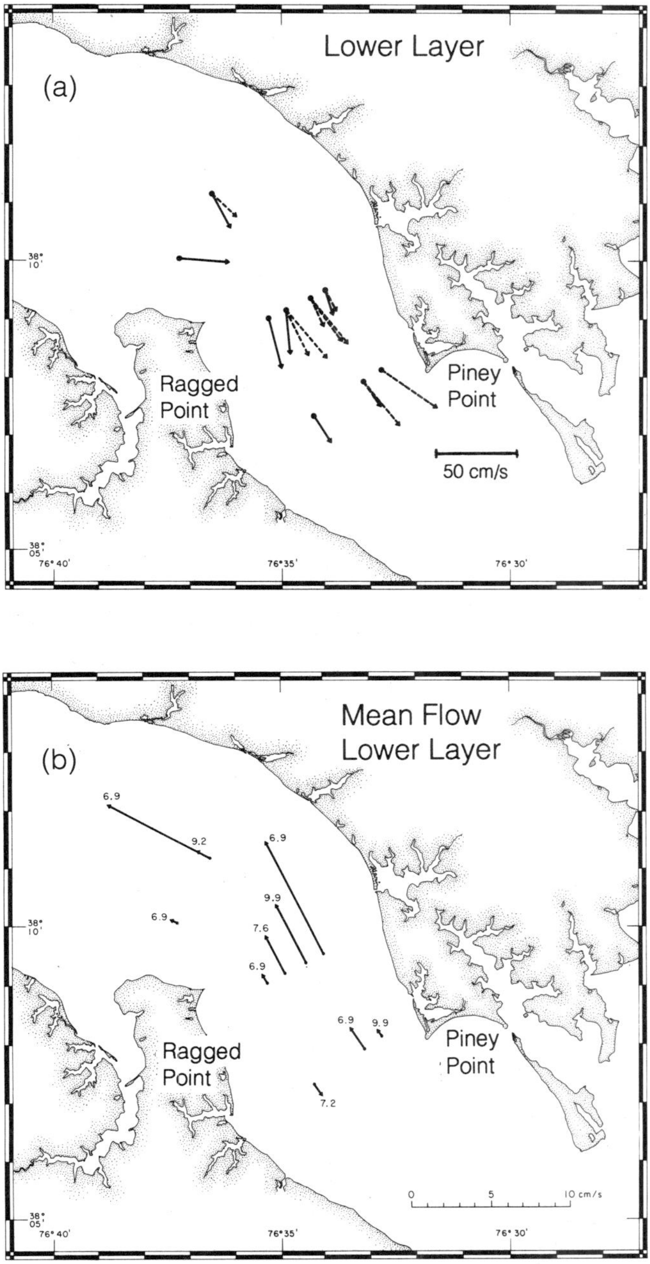


*Figure 5. (a) Instantaneous velocities at deeper levels of the Potomac River estuary at the same time as upper-layer velocities shown in Figure 4. (b) One-month mean velocities. Measurement depths (m) are indicated at the heads of the arrows.*

in the northern portion of the central channel. The pycnocline tilt may be sufficiently large that the depth of no net motion is deeper than the 7.2 m measurement on the southernmost mooring. At deeper depths, topographic interactions produce mean flows that are oppositely directed. The mean flow at 18 m in the central channel mooring off Ragged Point indicates a seaward flow of 2 cm/s, while the 18 m instrument in the channel mooring off Piney Point (4 km seaward) shows an upriver mean of 4 cm/s.

Bottom topography also influences the location of some estuarine fronts that are generated and destroyed on tidal time scales. The mechanism for the generation of longitudinal fronts is probably differential advection (with the faster flow over the center channel moving water of different density alongside the slower moving water in the adjacent shallows) in the presence of a longitudinal salinity gradient, as discussed by Huzzey and Brubaker (1988). Lateral fronts, also locked to bottom topography, can be formed in the lee of headlands. Kuo et al. (1988) described a James River front where an ebb-flow eddy recirculates saltier water back to the frontal region (locked to a steep bottom slope), whereupon it plunges under the lighter water on the opposite of the front. Estuarine fronts, whether tidally generated and topographically locked or large-scale features resulting from pycnocline surfacing (Beardsley and Boicourt 1981), are likely to be important to vertical and lateral exchanges, as pointed out by Garvine (1977). Estuarine plume fronts such as those emanating from Chesapeake Bay (Figure 6) (Boicourt et al. 1987) can occur within the Bay, as low-salinity water from tributary estuaries enters the main stem.

## *Lateral Processes — Upwelling, Tilting and Internal Seiching*

Most of our knowledge of estuarine circulation has been derived from field programs that have ignored (or deliberately eliminated through averaging) the variability in the cross-estuary, or lateral dimension. Although hints of potentially important lateral processes have been observed, the need to simplify the physics and to allot limited observational resources in a field effort usually resulted in experimental designs that favored the longitudinal and vertical di-

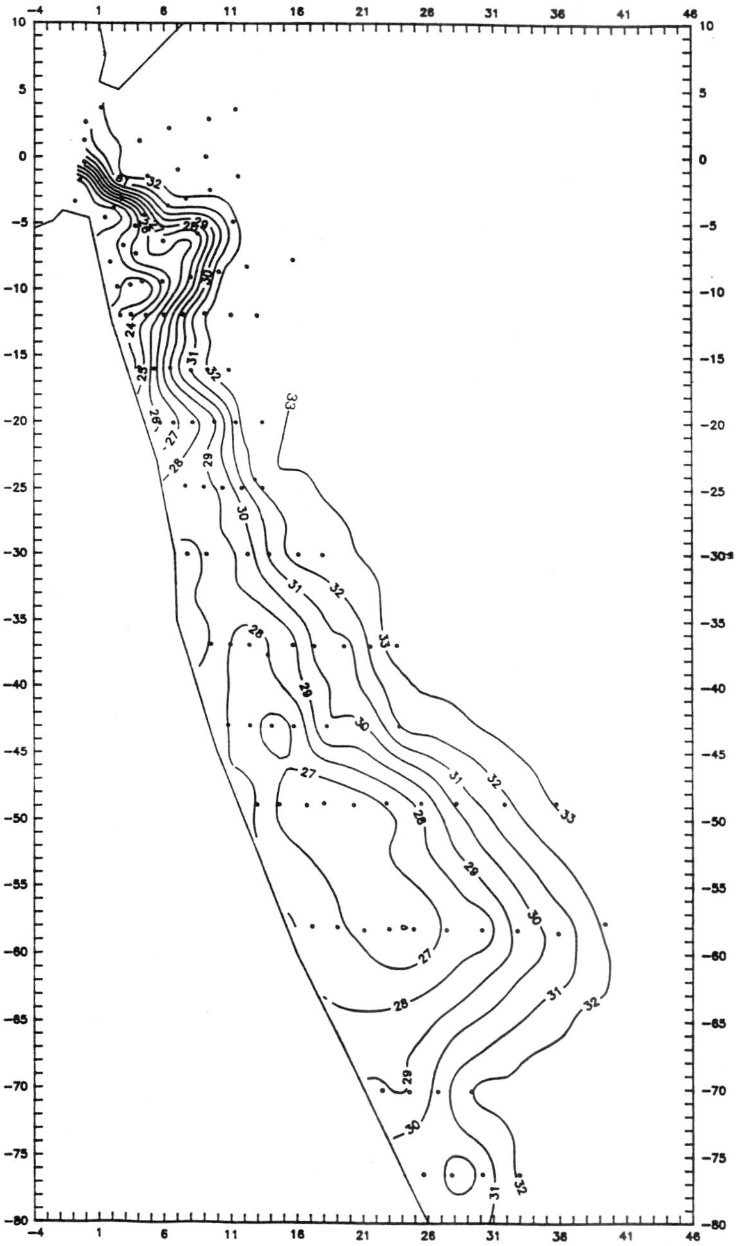


*Figure 6. Surface salinity map of the Chesapeake Bay plume on April 7, 1986. Initial turning region near the Bay entrance is bounded by a strong salinity front.*

mensions. In recent years, investigators have begun to explore the variability in the lateral dimension, and to examine the role of this variability in the transport and mixing processes. Stewart (1957) showed that Pritchard's (1956) lateral dynamic balance in the James River could be maintained by centrifugal forces derived from channel curvature, without any contribution from eddy friction. Dyer (1973, 1977) and Doyle and Wilson (1978) provide supporting evidence for the importance of centrifugal forces from observations in the Vellar and Hudson estuaries. Dyer (1977) offered some suggestions as to likely modes of lateral circulation in estuaries. Some of these modes, such as helical secondary circulations generated by the conservation of vorticity in shear flow around a channel bend, have firm observational support. Others, such as the double-cell lateral flow structure he predicts for strongly stratified estuaries, need further investigation.

In partially stratified coastal plain estuaries such as the Chesapeake Bay, the most obvious lateral structure is controlled by the rotation of the earth. In the northern hemisphere, the pycnocline tilts downward to the right of an observer facing seaward. This tilt, which is in approximate geostrophic (or more specifically, Margules) balance (where the pressure gradient associated with the tilt is balanced by the Coriolis force associated with the motions of the two layers) with the shear of the gravitational circulation, amounts to 3 to 5 m pycnocline height difference across the middle reaches of the Bay. Pritchard (1952) recognized that wind forcing could reverse this tilt and drive upwelled water on the western shore of the Bay. He invoked this process to explain the occurrence of low-oxygen water along the shallows on the western shore, creating crab "jubilees" where crabs moved into the shallows in great numbers in attempts to escape the undesirable conditions.

Carter et al. (1978) observed wind-driven variations in dissolved oxygen on the western shore of the Bay, off Calvert Cliffs. They measured continuous vertical profiles of temperature, salinity and dissolved oxygen by means of an innovative "elevator" sampling system mounted on rafts (Figure 7). In their multiple-regression models, wind impulse rather than wind stress was employed as the

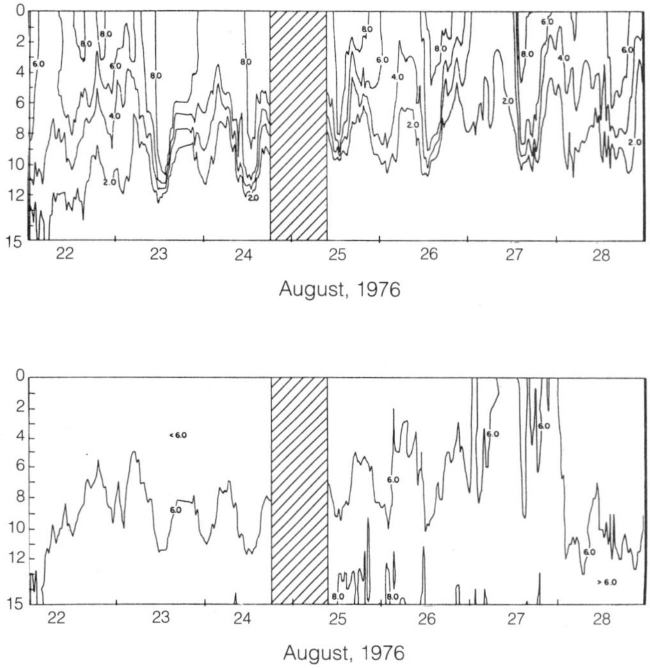


*Figure 7. Dissolved oxygen in ppm (upper) and density (sigma$_t$ ) (lower) off Calvert Cliffs during the week of August 22, 1976. Details are in Carter et al. (1978).*

independent variable for meteorological forcing. Wind impulse, which is wind stress acting over a period of time, produced significantly better regression models than wind stress. Part of the reason for the higher correlations between models and observations may have been the phase lag between forcing and response. The use of impulse will incorporate a phase lag that will vary according to the time history of the wind. Models were constructed for three depths: 0, 6 and 12 m. For the surface layers, the two major independent variables were offshore (cross-Bay) wind impulse, which accounted for 41% of the variance, and diurnal variation in solar energy (insolation), which accounted for 20% of the variance. At 12 m, the alongshore component of the wind impulse accounted for 38%, and the offshore component accounted for 10%. Both regressions had $r^2 > 0.8$. The conclusion from these regression models is that the most effective wind direction for driving a pycnocline tilt (and a low-oxygen event on the western shore of the Bay) is nearly parallel to the axis of the Bay. Lateral upwelling (Tyler 1984) and pycnocline tilting (Malone et al. 1986) have been further documented in the Chesapeake Bay. At present, however, there is a need to refine the picture of the time-dependent response of the pycnocline to wind driving.

A remaining question is the degree to which the pycnocline tilt is a forced, highly damped response and how much is a free oscillation of a lateral internal seiche. Higher-mode lateral internal seiches have been observed by Dyer (1982) in the Tamar estuary. We would expect that internal wave dynamics would exert a strong control on the response. The governing equation for the period T of a seiche of mode M in a two layer system with upper layer depth $h_1$ and lower layer thickness $h_2$, and b as estuary width, is:

$$T = \frac{2b}{MC} \qquad \text{(Eq 1)}$$

where C is the internal-wave phase speed:

$$c = \sqrt{g' \frac{h_1 h_2}{h_1 + h_2}} \qquad \text{(Eq 2)}$$

and g' is reduced gravity for the two-layer system with density difference ρ:

$$g' = \frac{\Delta\rho}{\rho_0} \qquad \text{(Eq 3)}$$

Support for both the forced-response and the internal-seiche view of the pycnocline tilt can be found in the subtidal motion of the Potomac River estuary. Boicourt's (1982) mooring array of current meters and temperature-salinity recorders showed pycnocline tilt reversals driven by southeasterly winds (Figure 8). During these events, the lateral velocity component was consistent with a tilting pycnocline. Low-pass filtering of these velocity and salinity records provides a picture of a forced response to the wind stress and a subsequent relaxation upon removal of forcing (Figure 9a). However, the unfiltered records (Figure 9b) reveal substantial oscillations, where salinity can fluctuate on the order of 6 salinity units over the tidal cycle. Longitudinal salinity gradients are insufficient to produce these oscillations by tidal advection.

Vertical motions of the pycnocline of tidal frequencies must be invoked to explain such signals. The decrease in the amplitude of these signals in the longitudinal direction supports the hypothesis that they are generated as internal lee waves at an abrupt change in bottom topography. As in the atmosphere, internal lee waves form in stratified flows downstream of a disturbance such as a mountain range, a submarine ridge, or a sudden increase in water depth. Such waves have been observed by Farmer and Smith (1978) in fjords and by Dyer and New (1986) and Brandt et al. (1986), in partially mixed estuaries. The combined velocity and salinity records from the Potomac study support the picture of a combined forced-oscillatory picture of the lateral pycnocline tilt, whereby the internal tidal oscillations are superimposed on a (forced) pycnocline tilt. In spite of the dense array of instruments in this study, spatial undersampling and concurrent energetic longitudinal processes prevented detailed views of the lateral pycnocline structure during the tilting events.

In the Chesapeake Bay proper, Sanford et al. (1990) observed both the forced upwelling response and lateral oscillations at tidal frequency off Calvert Cliffs. To first order, the internal tide appeared as an oscillation of amplitude 2 to 3 m superimposed on wind-forced motions of the pycnocline. Sanford et al. (1990) suggested that the internal tidal response would vary spatially because it appeared to be nearly resonant. Resonant conditions depend upon the cross-section geometry and stratification. Although spatial coverage and resolution of these measurements were also limited, phase information supported a picture of lateral first-mode oscillations. As in

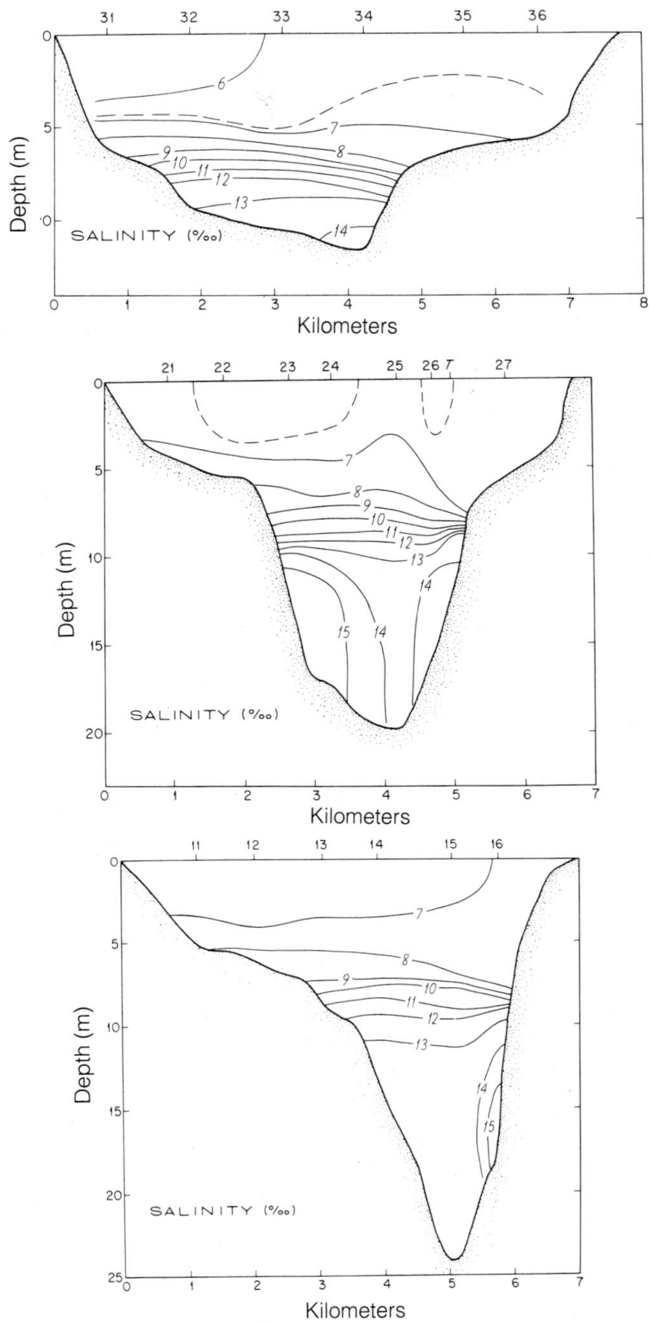


*Figure 8. Salinity (ppt) at the three cross sections containing the moored instrument array during the Potomac River experiment on June 13, 1978.*

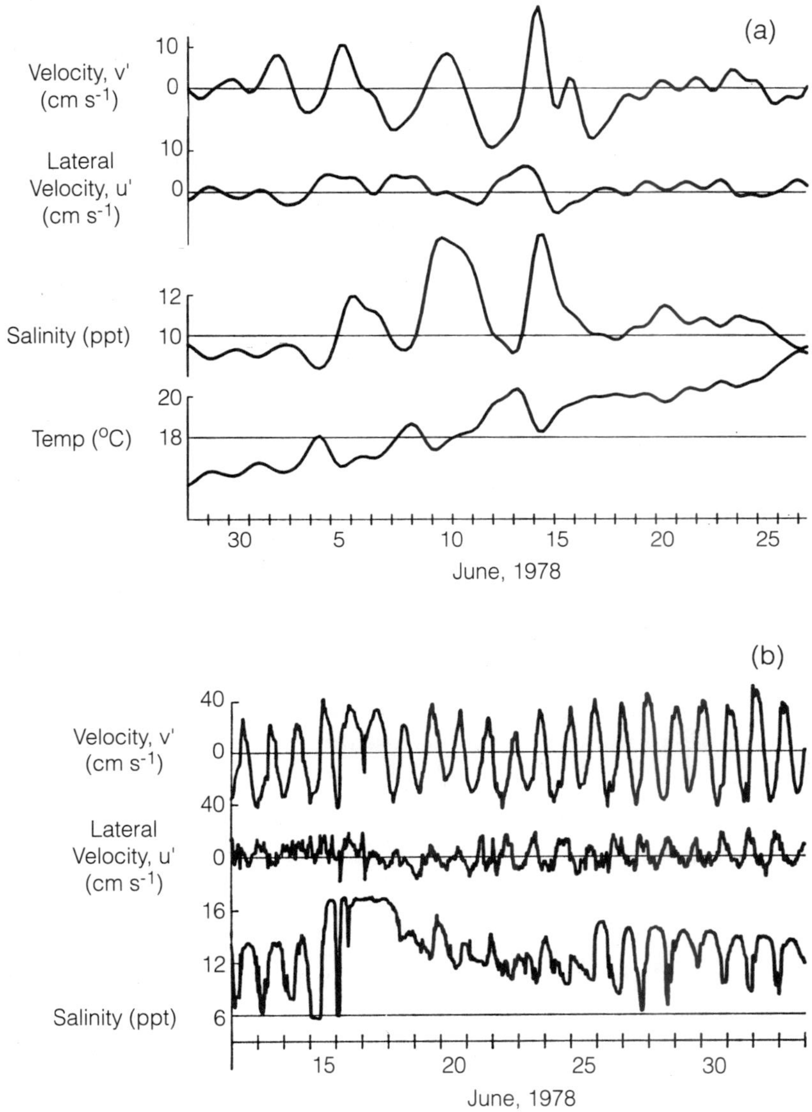


*Figure 9. (a) Low-passed filtered longitudinal velocity v', lateral velocity u', salinity and temperature at a depth of 9.9 m at mooring PM25 located in the central channel of the central array in the Potomac River experiment; (b) unfiltered portion of the record.*

the Potomac River case, more detailed coverage will be necessary to describe the pycnocline structure.

The increasing evidence for a myriad of lateral circulation processes (many of which provide structure and control on the longitudinal circulations) has triggered interest in the role they play in vertical exchange processes and, in turn, the effect on dissolved oxygen levels in the Bay's bottom waters. Although we understand that processes such as pycnocline tilting and lateral upwelling are likely to be important mechanisms of such exchange, we do not have any quantitative estimate of the transport associated with these processes. As a result, we do not know how to partition the exchange among the various component processes, or even a separation between lateral and cross-pycnocline exchange. The regularity of longitudinal salinity gradients in the upper and lower layers of the main stem of Chesapeake Bay might argue for quasi-steady mixing (such as turbulence generated by tidal currents interacting with the bottom topography) dominating the exchange processes. Possible exceptions might be major episodic wind-mixing events such as those documented by Goodrich et al. (1987), or sustained intervals of lateral upwelling along the flanks of the Bay. The problem remains, however, that processes traditionally relegated to the category of episodic events or spatially localized mixing (such as storms or pycnocline tilt reversals; Partch and Smith 1978; Brandt et al. 1986; Goodrich et al. 1987) may be the dominant modes of mixing and vertical exchange. Even a rough estimate of these small-scale processes is difficult because the quasi-steady processes of tidal mixing and gravitational advection act to smooth their effects.

### *Modeling Flow Dynamics*

Our observational and mathematical descriptions of the three-dimensional complexities of estuarine flows have evolved considerably from the classical two-layer, steady state model. Still, with all the evidence for intricate wind-driven and topographically induced flows, the major transport mechanism in Chesapeake Bay on time scales longer than the synoptic meteorological variability (greater than 7 to

12 days) is still gravitational circulation. Is our knowledge of gravitational circulation and wind-driven flows adequate to address the relationships between physical and biological processes? At present, the answer must remain equivocal. Until recently, no three-dimensional numerical model with sufficient spatial and temporal resolution had been developed for Chesapeake Bay. Such a model has been constructed and is presently being calibrated, verified and exercised on time scales appropriate to the oxygen-depletion problem (at least 5 months).

Modeling techniques, however, are not the primary limitation: greater obstacles to progress are the gaps in our understanding of the physical processes, especially with regard to the details of transport and dispersion. Such details of the vertical transport, or mixing, of oxygen are of vital importance to the quantitative description of oxygen depletion.

A revealing example of our lack of knowledge about mixing processes is the uncertainty of the mixing source for the three-layer circulation in the Baltimore Harbor-Patapsco River estuary. The cross-sectional area of Baltimore Harbor is large enough that, for most of the channel, the tidal currents necessary to fill and empty the intertidal volume (the volume of water between high and low tide, often called the tidal prism) are weak, with amplitudes less than 5 cm $s^{-1}$ (Boicourt and Olson 1982). These small tidal currents cannot produce the strong mixing necessary to drive the three-layer circulation via boundary mixing. While wind must be the primary source of mixing energy in this tributary, the particular mechanisms are uncertain.

Until our knowledge of circulation dynamics is at such a level that the gravitational, wind-driven and topographically induced flows can be accurately integrated within a numerical model, we will not be able to provide an accurate description of the influences of physical circulation on biological and chemical processes in Chesapeake Bay. Such a model would necessarily incorporate temporal and spatial details of the mixing processes. There should be optimism, however, that while we are developing this detailed description, empirical quantitative relationships such as the bulk Richardson number shown for Bay stratification by Goodrich et al. (1987) will help us decipher the primary connections.

Clearly, in the last decade there has been a substantial increase in our understanding of estuarine circulation processes — whether the present understanding is adequate for dealing with the problem of oxygen depletion is probably best answered by examining the role of physical processes in controlling long-term hypoxia trends.

## Hypoxia Trends: 1950-1980

In 1948, the Chesapeake Bay Institute (CBI) initiated a series of Bay-wide surveys of temperature, salinity and dissolved oxygen. This measurement series, with its consistent procedure for determining oxygen concentrations, was to provide major evidence that the Bay's water quality — as indicated by hypoxia — was in decline. Flemer et al. (1983) estimated the volume of summertime oxygen depletion from these data and emphasized the dramatic difference in the volume of anoxia between the years 1950 and 1980. Using the same data, Officer et al. (1984) argued that the increase of oxygen depletion over time is related to the yearly increase of nutrient inputs to the Bay and the consequent rise in annual plankton production. One of the primary arguments for a nutrient-produced increase in summer hypoxia rather than a stratification-produced increase is the lack of a concurrent (1950 to 1980) increase in fresh water inflow to the Bay. When EPA reported the results of Flemer et al. (1983), scientists argued that the 1950 to 1980 trend could not be supported statistically by a record with as much interannual variability. At the time, a river-flow normalization was attempted, but the correlations were so low that the connection was not established.

Seliger and Boggs (1988) reexamined the CBI oxygen data and recalculated the volumes of hypoxic water in the interval 1950 to 1985 (Figure 10). They found, in contrast to the conclusions of Flemer et al. (1983), that the volume of hypoxic water was related to the April to May Susquehanna River flow, with an $r^2$ of 0.83

(Figure 11). The discrepancy in conclusions derived from approximately the same data is substantial, with each having fundamentally different implications to the Bay's ecosystem. Although the interannual variability is large, the calculations by Flemer et al. (1983) appear to indicate an increasing anoxic volume over time; in the calculations by Seliger and Boggs (1988), the trend is less obvious. The $r^2$ between the hypoxic volumes determined by Flemer et al. and the April to May Susquehanna River flow is only 0.08 (Figure 12).

Why, when using the same data and the same geometrical information on Chesapeake Bay (Cronin and Pritchard 1975), is there such a discrepancy? To begin with, each estimate employed slightly different threshold values for hypoxia. More importantly, different data-selection criteria were used. Seliger and Boggs included the years after 1980, but they also included the years 1952 and 1958, which were excluded by Flemer et al. (1983, Appendix B, Section 5). Both 1952 and 1958 were years of extensive anoxia in the Bay, with 1958 appearing as the most widespread on record. The removal of these two years would reduce the correlation determined by Seliger and Boggs and, conversely, the inclusion would increase those of Flemer et al. (1983). Further support for the connection between river flow and oxygen depletion is provided by Malone (this volume), whose time-integrated thickness of hypoxic and anoxic waters (although not a direct measure of depletion volume) is related to river flow in the years 1984 to 1988.

The suggestion for the primacy of physical process control of dissolved oxygen on at least weekly time scales comes from the demonstration by Taft (1983) that oxygen differences across the pycnocline are related to density differences and, hence, stratification. He assembled all available July data from the two CBI stations located in the region where the maximum oxygen depletion occurs and found that these differences in dissolved oxygen and density were related with an $r^2$ of 0.76 (Figure 13). The argument, then, is this: because stratification controls vertical oxygen profiles on both short and long time scales, stratification plays the primary role in controlling oxygen depletion in the deep layers of the Bay. However, the relationship is not exact, and there is a possibility that the amount and timing of nutrient delivery plays a significant role in controlling the volume of anoxic water. Taft (1983) showed that stratification control of the vertical oxygen gradient does not necessarily operate during the spring, when anoxia is developing. A possible support for a trend in the anoxia is that the May 1950 data show stratification control, while the May 1980 data do not. It is important to note that stratification does not necessarily control the absolute concentrations, but the gradient or concentration difference across the pycnocline. If there has been a recent downward trend in upper-layer oxygen values, then increased nutrient loading would be a prime candidate for the cause of such a shift.

While spatial undersampling may explain the difference in calculations of anoxic volume, a more serious problem is the result

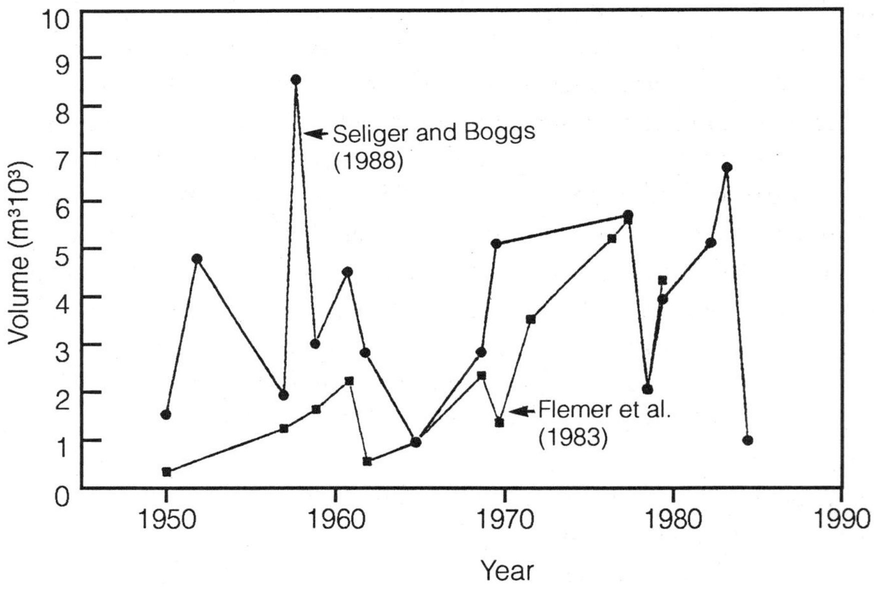


*Figure 10. Volume of anoxic water in summer for Chesapeake Bay. Shown are volume of water less than 0.5 ml L$^{-1}$ (Flemer et al. 1983) and volume of water less than 1 ppm (Seliger and Boggs 1988).*

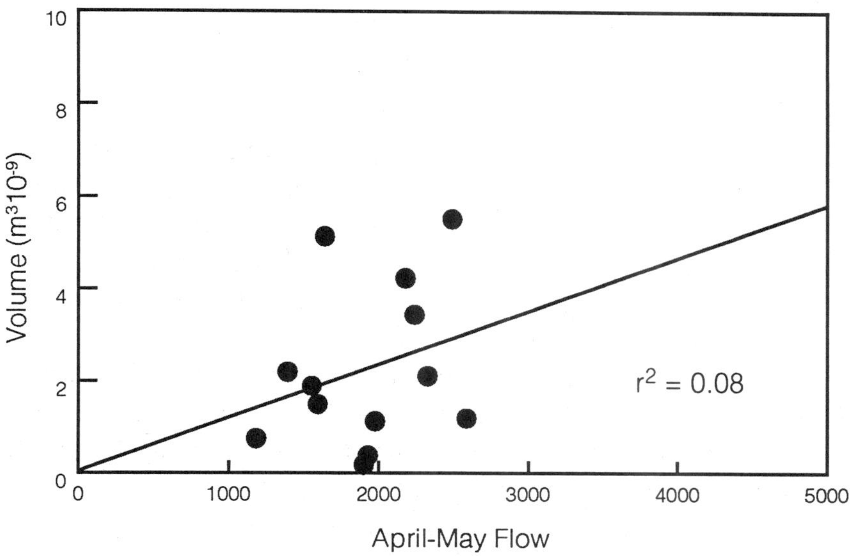


*Figure 11. Volume of anoxic water in Chesapeake Bay versus April-May freshwater flow from the Susquehanna River (data from Seliger and Boggs 1988).*

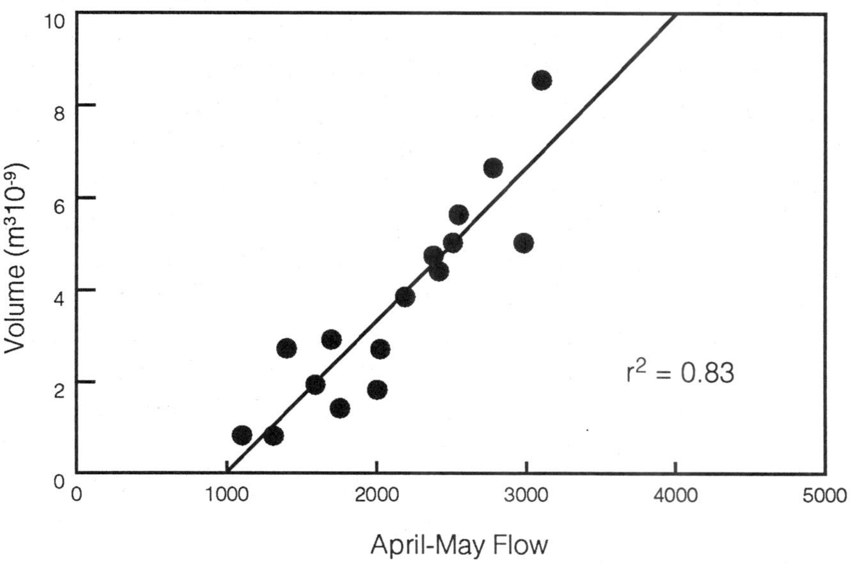


*Figure 12. Volume of summer anoxic water in Chesapeake Bay versus April-May freshwater flow from the Susquehanna River (data from Flemer et al. 1983).*

of temporal undersampling. The high correlations found between spring runoff and summer oxygen depletion were determined from oxygen measurements made only once or twice during the summer. Malone et al. (1986) observed short-term fluctuations in dissolved oxygen that he ascribed to meteorologically driven processes such as lateral tilting of the pycnocline and associated upwelling or enhanced vertical mixing during storms.

Prior to yet another examination of the same CBI data set (which is being undertaken at the Horn Point Environmental Laboratory), two observations can be made concerning the anoxic volumes: (l) given the sometimes sparse data set and the geometry of Chesapeake-Bay cross sections, the procedure for determining hypoxic volumes may produce substantial uncertainty; (2) the April to May freshwater flow can explain a significant portion of the interannual variability in oxygen depletion, and the connection appears to result from the suppression of vertical exchange by stratification.

Another conclusion from the river-flow normalization of the anoxic-volume variability is that the time-series record cannot be used as *prima facie* evidence for an increasing trend in oxygen depletion spatially and temporally. Though this record was not the only evidence cited by the Environmental Protection Agency (1983) for such a trend, it was a major component of the case. It is important that scientists and managers be cautious in the conclusions drawn from the indicated river-flow normalization of the hypoxia record. This normalization reduced the trend inferred by EPA to the level of statistical insignificance; on the other hand, this normalization is not sufficiently tight that it precludes the existence of a trend, which may be slight, although it is symptomatic of significant changes in the ecosystem. The normalization does indicate that careful, detailed and accurate process observation and modeling will be required to definitively detect or disprove the existence of such a trend in the extant records.

Given the magnitude and expense of the planned nutrient-reduction efforts on Chesapeake Bay, establishing whether or not there is such a trend should be a research focus. The existence of a

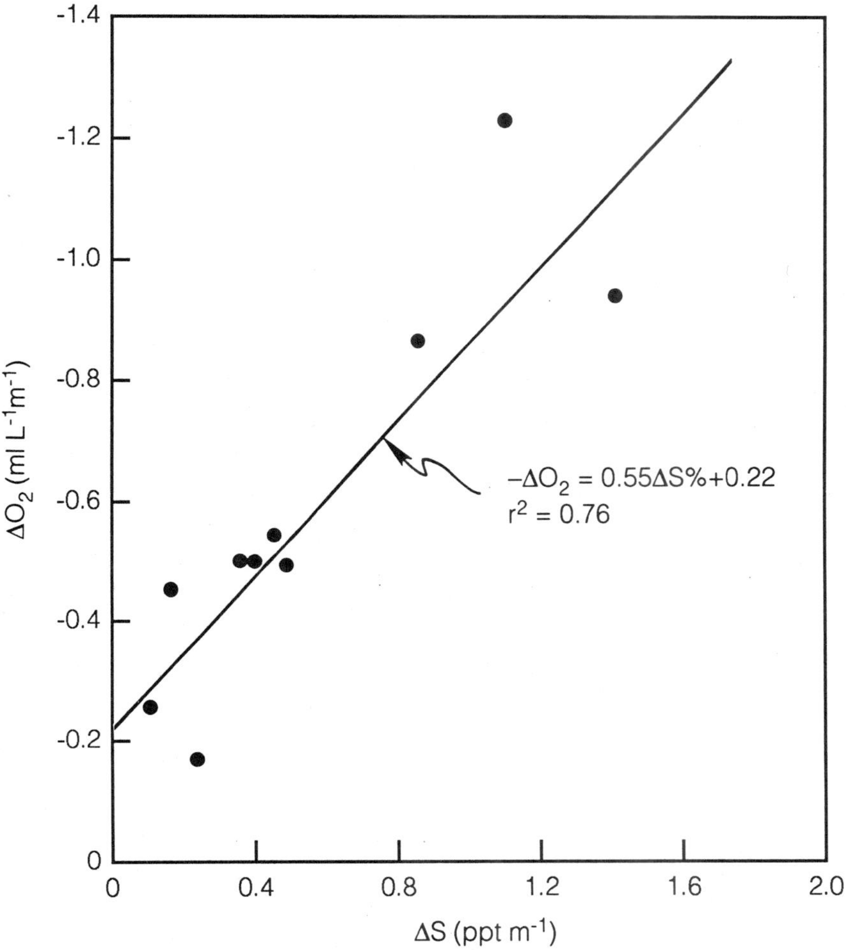


*Figure 13. Oxygen differences across the pycnocline versus salinity differences for mid-Chesapeake Bay stations, 1949-1980 (Flemer et al. 1983).*

trend, however, should not constitute the definitive test for the adequacy or advisability of specific nutrient-reduction strategies for the Bay. There is enough evidence for the deleterious effects of excess nutrient loading to the Bay on other grounds so that nutrient reduction would appear beneficial, for instance, for reversing the widespread loss of submerged aquatic vegetation. But the lack of a trend in anoxic volume may signal that it should not be the sole indicator by which to gauge the health of Chesapeake Bay.

## Major Influence on Hypoxia: Freshwater Flow or Nutrient Input

Although the strength of the relationship between spring runoff and summer anoxic volume has not been definitively determined, the connection has been well established (Seliger and Boggs 1988; Malone, this volume) so that we would like to examine the controlling mechanisms of oxygen depletion in the Bay's bottom waters. Spring runoff delivers the major annual supply of fresh water (Schubel and Pritchard 1986) and nutrients (Malone, this volume). Because river flow increases stratification, and thereby suppresses vertical exchange, we need to distinguish the relative importance of each process on a seasonal time scale. Because the processes of nutrient delivery and stratification are coupled, simple correlative connections will not suffice to separate their influences.

One difficulty with the hypothesis that spring river flow/stratification controls hypoxia concerns the phase relationships. For the middle reaches, or mesohaline region, of Chesapeake Bay, the phase lag between peak flow of the Susquehanna River and peak stratification is on the order of one month. The phase lag between the spring river flow and the summer oxygen depletion, however, is 2 to 3 months. If spring river flow controls the stratification at one-month scales, it would seem unlikely to control vertical exchanges over the longer time scales, and therefore spring runoff may not be controlling the anoxia during summer.

An examination of salinity records from a long-term mooring in Chesapeake Bay off the mouth of the Patuxent River will help illustrate this apparent dilemma. The relationships between peak flows of the Susquehanna River and the vertical salinity differences at the Patuxent mooring seem bimodal (Figure 14). The primary response in the stratification appears at the one-month time scale. There is also some suggestion of a secondary, lower-amplitude response shortly after the peak flows of the Susquehanna. These rapid increases in stratification are probably the result of upper-layer intrusions of fresh water propagating down the Bay in a manner similar to the bore intrusions observed in the Chesapeake Bay plume on the continental shelf (Chao and Boicourt, 1986). Because they only occur as a one-layer gravity current, they only affect the upper-layer salinity.

The primary, one-month response of mid-Bay stratification to a freshet from the Susquehanna River is a consequence of the freshet's spur to the gravitational circulation. The accelerated lower-layer flow toward the head of the Bay increases the landward transport of salt. Such a "rebound" response can result in a freshet *increasing* the salinity of lower-layer waters in the upper reaches of the Bay over levels prior to the freshet. Although observational lore provides examples of this rebound process (Schubel and Pritchard, 1986), the details of such transient and time-dependent responses to river-flow pulses are poorly known. The influence of topographic controls in these (upper- and lower-layer) longitudinal intrusions is strongly suspected (Farmer and Armi 1986; Chao and Paluszkiewicz 1991), but the process and its effects have not been adequately observed in the Bay.

A possible resolution of the apparent discrepancy in time scale between the one-month peak-to-peak response time and spring runoff control of summer anoxia can be found by examining a strong wind-mixing event, which might be expected to decouple this control. In mid-June of 1986, for example, a storm destratified Chesapeake Bay in the manner described by Goodrich et al. (1987). This event can be seen in Figure 14 as a rapid decrease in lower-layer salinity at the long-term mooring, and a resulting vertical salinity difference approaching zero. It could be argued that in the absence of any appreciable river flow within the month preceding, this event should effectively decouple the stratification (and hence, the hypoxia) from any control that the spring runoff might exert. But in early July, 1986 (Figure 14), the Bay restratified without a noticeable driving event in the flow of the Susquehanna River. The buoyancy to supply this restratification came from the *horizontal* salinity gradient along the axis of the Bay. Although destratification mixed the water column to near homogeneity, upper Chesapeake Bay remained fresher than lower Chesapeake Bay, and the Bay as a whole remained fresher than the adjacent continental shelf waters. This salinity gradient was then available to drive a gravitational flow of lower salinity water moving seaward over landward-moving higher salinity water, thereby restratifying the water column.

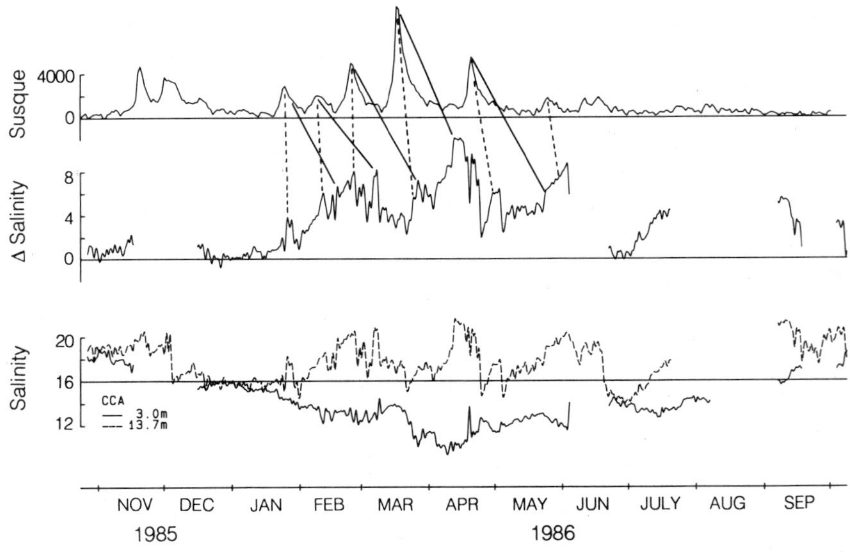


*Figure 14. Susquehanna River flow and salinity for mid-Chesapeake Bay, 1985-1986. Salinity was measured at station CCA off the mouth of the Patuxent River at a depth of 3 m (solid line) and 13.7 m (dashed line). Bottom panel shows salinity difference between upper and lower layers. Solid and dashed straight lines are drawn between river flow events and possible corresponding events in the salinity-difference signal.*

This ability of the Chesapeake Bay to restratify after wind-mixing events depends upon the longitudinal gradient of cross-sectional mean salinity along the axis of the Bay, which in turn depends upon the prior history of freshwater runoff (Boicourt 1969). For the Bay as a whole, this ability would presumably be at a maximum after spring river flow, when the mean salinity of Chesapeake Bay is at a minimum. The springtime buildup of fresh water within the Bay can be seen as a "buoyancy reservoir" that is later available for restratification after wind mixing. This buoyancy reservoir provides a mechanism for the runoff maximum to exert an influence on stratification that can extend through destratification events to the summer hypoxia. Although the 2 to 3 month interval between maximum runoff and summer hypoxia is longer than the observed one-month delay between Susquehanna River freshets and stratification peaks, spring runoff can determine summer stratification by setting the starting point for a gradual decay as river flow decreases.

The spring river flow therefore can control the summer stratification, and the volume of anoxic water, through the buildup of fresh water — a buoyancy reservoir — in the Bay. The time scale for the effects of this buildup will be greater than the one month peak-to-peak response time. A possible explanation for the Seliger and Boggs (1988) high correlation between river flow and oxygen depletion from temporally undersampled data may lie in the ability for the Bay's buoyancy reservoir to reestablish an equilibrium stratification after short-term meteorologically driven reaeration events (Malone et al. 1986). With high rates of biochemical oxygen consumption in bottom waters, the hypoxia could then be reestablished to levels controlled by the stratification. Since the CBI cruises were usually conducted over intervals of approximately one week, the oxygen-depletion volumes may be relatively unaffected by the occasional short-term intervals of oxygen reaeration. Such a possibility can be tested by a reexamination of the higher-frequency oxygen data of Malone et al. (1986) and Sanford et al. (1990), in conjunction with the spatially extensive data of the Chesapeake Bay Monitoring Program.

Although the evidence supports the conclusion that stratification influences the volume of hypoxic or anoxic water in Chesapeake Bay, the question remains: how can this physical process serve as the primary regulatory mechanism for the fundamentally biochemical process of oxygen depletion? The first part of the answer is that stratification does not control oxygen depletion *per se* but the rate of vertical exchange and, therefore, the rate of reaeration of lower-layer waters. The second part of the answer is a finding (Malone, this volume) that the rate of oxygen consumption in the Bay's bottom waters is relatively invariant from year to year. Therefore, the balance between oxygen consumption in waters below the pycnocline and supply of oxygen from above appears to be controlled by variations in supply, which is in turn controlled by stratification.

A difficulty with this simple explanation of oxygen depletion is that water column oxygen consumption and sediment oxygen consumption are combined as a net consumption. Kemp and Boynton (this volume) show that sediment oxygen consumption appears to have greater interannual variation than oxygen consumption in the lower-layer water. In addition, they demonstrate that benthic nutrient retention and recycling produce an oxygen-demand component with a one-year time lag. Their conclusion that sediment oxygen consumption is important only at the beginning and the end of the summertime oxygen depletion may provide an explanation for possible water-column dominance in oxygen consumption. However, separation of these effects will be necessary to remove uncertainty in the simple stratification-control model.

Although we can explain how spring runoff may control summertime stratification in Chesapeake Bay, it is not yet possible to provide a quantitative description of the process. One reason for this difficulty is the influence of wind-driven mixing variability, not only in generating noise in the stratification record, but also in generating mean effects through nonlinearities. Simple arguments have supported the strong influence of river flow on hypoxia and anoxia volume through stratification; however, substantial work remains before we can clearly demonstrate the relative roles of stratification and nutrient delivery in producing the degree of summertime oxygen depletion.

## Physical Processes and Biological Rates

Within a given water body, the present distribution of properties, whether conservative (not created or lost) or non-conservative, reflects the processes of advection and dispersion integrated over some time prior to the present. If we know some of these processes or property distributions, then we can use this knowledge to extract insight into the other, lesser-known processes. Temperature and salinity, for instance, serve as semi-independent tests of advection and dispersion models. Although salinity is not strictly conservative in the estuary, it is nearly so and therefore an especially valuable tracer.

Temperature is less conservative than salinity in the estuary, but it does not warrant the traditional neglect as a tracer of estuarine circulation. Seitz (1971) demonstrated the value of temperature information in Chesapeake Bay. From monthly hydrographic transects up the axis of the Bay, he revealed the seasonal development of a temperature minimum in the vertical. He also revealed a temperature minimum in the horizontal, at the northern end of the deep channel near the Bay Bridge (Figure 15). This deep minimum does not imply a static mass of cold water trapped at the end of the deep channel. Current measurements and salinity distributions indicate that there is a steady upestuary flow through this minimum. The minimum can be explained in its simplest form as a dynamic balance between horizontal advection and vertical mixing. As the lower layer gradually warms in the spring and summer, the estuarine circulation brings warmer salty water from the southern reaches of the estuary and carries it under the slower-moving water near the pycnocline. The increase in lower-layer temperature northward of the minimum is the result of shoaling of the channel and an associated increase in vertical exchange (both advection and diffusion).

The same processes that determine the structure and location of the temperature minimum in Chesapeake Bay also control the distribution of oxygen. For this reason, the oxygen-depletion zone has a close correspondence to the temperature minimum. Over the southern part of the zone, vertical oxygen profiles often show a mid-depth minimum. In addition, the horizontal minimum in oxygen is usually at the northern end of the deep channel. As in the case of temperature, this distribution structure is, to first order, controlled by a balance between vertical exchange and horizontal advection. We can write a simple balance for the lower layer dissolved oxygen, following a parcel of water:

$$\frac{DO_2}{Dt} = \frac{\partial O_2}{\partial t} + v\frac{\partial O_2}{\partial x} = Q + S_i \qquad \text{(Eq 4)}$$

where:

> v is the velocity component in the longitudinal or y axis, with positive flow directed out of the estuary
>
> Q is the source term, which includes both internal sources and vertical exchange from the upper layer
>
> $S_i$ is consumption from both the water column and the bottom sediments

Observed distributions can be employed to estimate these nonconservative rates of oxygen consumption. Two successive longitudinal profiles of surface and bottom dissolved oxygen from the Chesapeake Bay Program Mainstem Monitoring are shown in Figure 16. From the perspective of a water parcel moving up the estuary, the decline in near-bottom oxygen represents the consumption by benthic and water-column processes along its path. This decline in oxygen is remarkably linear from the entrance to the Bay to the oxygen minimum zone. Some of the indicated decline is the result of bottom depth increasing from the lower Bay to the deep channel that begins below the mouth of the Potomac River. The depth of this deep trough does not vary significantly from the Potomac River to the Bay Bridge.

From the view of an observer at one location in the estuary, the upestuary flow in the lower layer represents an influx of higher-oxygen water. If we approximate the longitudinal oxygen gradient as linear, then we can estimate a lower bound for the rate of consumption in the lower layer. Given that the horizontal deep water velocity v is on the order of −10 cm $s^{-1}$, then the advective contribution to the oxygen balance in the lower layer is 0.4 mg $L^{-1}d^{-1}$. If we assume that there is no source of oxygen from vertical exchange or from photosynthesis, then the magnitude of the consumptive rates must be at least 0.4 mg $L^{-1}d^{-1}$ to balance the contribution from horizontal advection. If these benthic and water-column rates of consumption did not surpass advective reaeration, then the oxygen-depletion zone would disappear. During the two-week interval between the two cruises (Figure 16), oxygen levels increased in the southern portion of the Bay, below latitude 38° 40' N. This increase is probably a combination of both advection and vertical exchange, because oxygen levels increased in the upper layers during the same interval. The rate of lower-layer oxygen increase is less than the rate of advective reaeration, indicating that consumption is still active.

To maintain or decrease oxygen levels in the Bay, consumption must occur at sufficiently high rates to balance or surpass reaeration from both horizontal and vertical exchange. What, then, is the controlling reaeration process? Kuo and Neilson (1987) show that variations in the strength of the gravitational circulation can explain differences in hypoxia between the James, York and Rappahannock rivers in the southern part of Chesapeake Bay. Here, stronger gravitational flows provide increased oxygen supply or a decreased transit time through the deeper hypoxic regions of the tributaries. In the main stem of Chesapeake Bay, increased spring river flow increases both the gravitational circulation and stratification. However, gravitational circulation would act to increase dissolved oxygen concentrations in the lower layers of the Bay (by advecting water with higher concentrations of oxygen from the southern portions of the Bay), while stratification acts to decrease oxygen levels. The correlation between spring river flow and oxygen-depleted volume indicates that the stratification control wins the competition between these two processes. Although stratification ap-

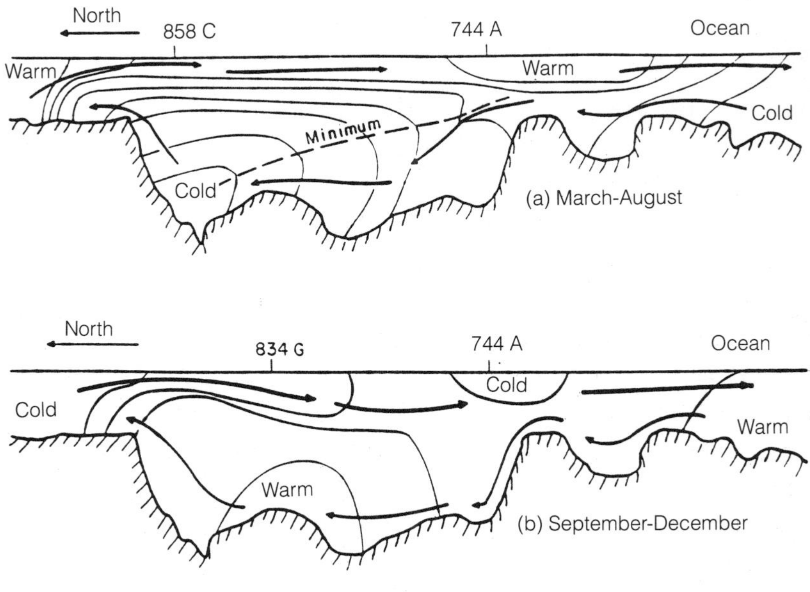


*Figure 15. Schematic diagram of the development of (a) the deep Chesapeake Bay temperature minimum during March-August and (b) the development of the temperature maximum during autumn (from Seitz 1971).*

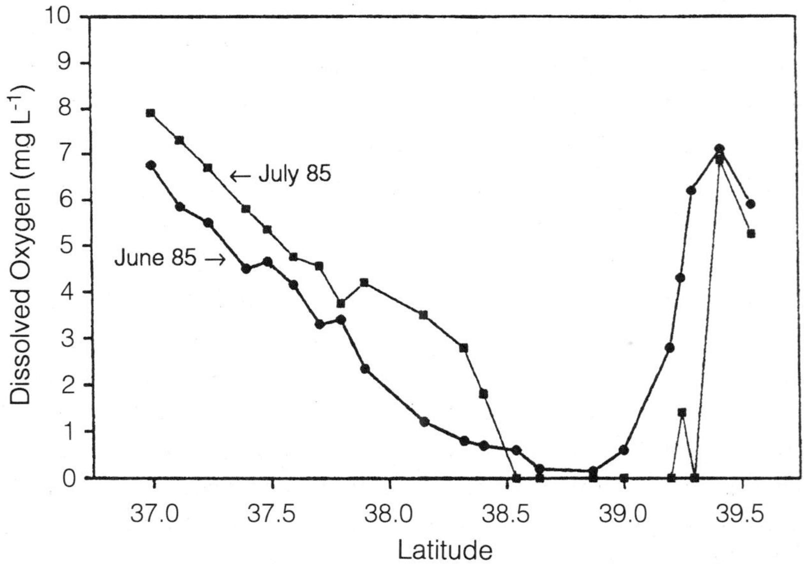


*Figure 16. Longitudinal distributions of near-bottom dissolved oxygen in Chesapeake Bay for two successive cruises during June 17-19, 1985 and July 8-10, 1985 (data from the EPA Chesapeake Bay Program).*

pears to win over advection in the main stem, the demonstrated effects of gravitational circulation in determining the shape and location of the oxygen-depletion zone imply that advection is far from negligible. These processes are highly coupled and, therefore, deciphering the process details will not be as simple as the linear connection between runoff and oxygen-depletion volume might suggest.

In summary, knowledge of the physical processes of advection and diffusion can be employed to develop estimates of net rates of consumption. Temperature and oxygen have a parallel structure during the spring and summer seasons in Chesapeake Bay because the sources for both heat and oxygen are located nearly uniformly at the Bay's surface and in the southern reaches of the Bay. In addition, the oxygen sink is located in the Bay's deeper layers (and at the bottom), where cooler temperatures remain from the previous winter's cooling. Vertical exchange will transport heat and oxygen from the upper layer to the lower layer, while the gravitational circulation will bring warmer, more oxygenated water from the shallow regions in the southern Bay. The vertical velocity profile of the gravitational circulation is such that this warm water will underride slower-moving water and create a temperature and oxygen-minimum layer immediately below the pycnocline.

Topography and advection confine the oxygen-depletion zone to the northern portion of the deep channel of the Bay. The deep channel is so well insulated from vertical exchange that consumption can overcome supply from production and reaeration. As with the temperature minimum, the stationary position of the oxygen minimum should not imply that the water is also stationary. Both the temperature and oxygen minima are dynamic features, with inflows and outflows. As a parcel of water moves up the Bay in the lower layer, consumptive processes act to remove oxygen. At the north end of the deep channel (near the Bay Bridge), the parcel is moved to shallower depths by the gravitational flow. In the shallower depths north of the Bay Bridge, reaeration by vertical exchange overcomes consumption and the oxygen concentration of the water parcel increases. The topographic shoaling at the north end of the deep channel usually provides a strong constraint to the northward propagation of hypoxic or anoxic waters. In years of extensive anoxia, or during strong up-Bay surges driven by meteorological forcing, oxygen-depleted water can be found north of the deep channel.

## Learning from the Tributaries

The observations of summer oxygen depletion in Chesapeake Bay tributaries has raised questions about whether such occurrences are controlled by local processes, or are imported from the main stem of Chesapeake Bay. The implications for management are obvious: if local processes determine the health of the tributary, then actions necessary for restoration and conservation of water quality would be limited to the domain of the tributary and its watershed. If, on the other hand, the primary control of oxygen depletion derives from the Bay proper, then the magnitude of the management problem becomes substantially greater.

In spite of the additional complexity provided by the Bay-versus-local-control question, recent studies designed to address the water quality in tributaries can offer helpful insight to dissolved oxygen processes. The smaller scales of tributaries make them more amenable to high-resolution observation. More importantly, they represent microcosms with similar, but somewhat independent processes from which to gain insight into the dynamic balances of oxygen and nutrients in the coupled Bay-tributary system. The Patapsco River estuary (Baltimore Harbor) is an example of such a microcosm, where the main stem provides the density structure (and the potential energy) for three-layer gravitational flow, though local mixing processes within the tributary exert the primary control on the strength of the circulation (Boicourt and Olson 1982).

Kuo and Neilson (1987) provide an explanation for the paradoxical variation in hypoxia in the Bay's Virginia tributaries. The James River receives a greater nutrient loading than the York River or the Rappahannock River and yet has the fewest episodes of hypoxia. Kuo and Neilson (1987) suggest that the stronger gravitational circulation of the James River minimizes the transit time of water through the deep hypoxic region of the estuary. A stronger gravitational circulation would also provide reaeration of water from the lower reaches of the tributary, or outside the tributary, from the Bay proper. An alternate explanation might be that, given the longitudinal gradient in dissolved oxygen in the Bay (Figure 16), the northward-increasing hypoxia in the tributaries is the result of preconditioning of the source water from the Bay, where bottom oxygen concentrations decrease in a northward direction. Kuo and Neilson (1987) point out that some of their data indicate this possibility. While they think that the transit-time differences between the James and the York rivers is the most likely explanation, they also feel that the influence of source-water explanation might affect the difference between the York and the Rappahannock rivers.

Recent studies of the Patuxent and Choptank rivers were motivated by water quality concerns. Increased nutrient loading to the Patuxent watershed has been correlated with an increase in oxygen depletion within the estuary (Heinle et al. 1980). The understanding that we now have from recent studies is that local oxygen consumption within the middle reaches of the estuary exerts the primary control on low-oxygen levels (Domotor et al. 1989). Oxygen depletion in the bottom waters of the Patuxent is also enhanced by intrusions of anoxic waters from the Chesapeake Bay (Boicourt and Sanford 1990). These intrusions are generated by two different wind-driven mechanisms that tilt the Bay pycnocline and draw anoxic water over the entrance sill. Once this high-salinity water surmounts the sill, it propagates up the Patuxent River as an internal bore along the bottom. These intrusions appear as surges, superimposed upon the lower-layer inflow of the steadier estuarine circulation.

In the Choptank, intrusions of anoxic water from the main stem of Chesapeake Bay were thought to adversely affect oyster populations (Seliger et al. 1985). However, Sanford and Boicourt (1990) found that these intrusions rarely penetrate to active oyster bars. They documented intrusions which were set up by the lateral tilting of the pycnocline, and which were controlled by a series of sills along the Choptank entrance channel.

In both the Patuxent and Choptank rivers, the occurrence of oxygen depletion in apparently isolated basins in the middle reaches of the tributary (Domotor et al. 1989; Sanford and Boicourt 1990) suggests the dominance of local consumptive processes over advection from the Bay's main stem. In addition, the phase lead in oxygen depletion of these basins over depletion in the Bay proper (Domotor et al. 1989) adds further support for local control over main stem remote control. In such dynamic, coupled systems, however, achieving a quantitative partitioning among the component processes of local consumption and production, vertical exchange, and horizontal advection will require detailed information on each component. Until this information is incorporated into a numerical model, we will not be able to resolve such questions as why, for example, Broomes Island is a locus for oxygen depletion in the Patuxent River (Domotor et al. 1989). We will also not be able to properly characterize the role of intrusions of lower-layer water from the Chesapeake Bay. The possibility exists for these intrusions to provide a trigger mechanism for oxygen depletion, even though the primary consumption operates locally.

In addition to the larger tributaries such as the Patuxent and Choptank rivers, smaller tributaries such as Rock Creek off the Patapsco River (Copp and Boicourt 1988) provide test cases for the study of oxygen depletion. When these cases are combined with the larger tributaries, they form an ensemble within which each estuary is driven by approximately the same seasonal signal of runoff, nutrient loading and atmospheric forcing. In this context, the smaller deviations of the signals from the means and the individual behavior of each tributary can be revealing. For this collective view to be effective, quantitative knowledge of the exchanges between the tributary and the main stem is essential. The ongoing monitoring programs of the State of Maryland, the Commonwealth of Virginia and the U.S. Environmental Protection Agency are a valuable source of information from which to estimate end-member concentrations for these exchanges.

## Research Summary and Management Needs

We have seen that the structure and circulation of Chesapeake Bay waters are primary influences on dissolved oxygen processes. However, the present explanation that stratification controls the balance between oxygen supply and consumption in the lower-layer waters of the Bay is only a first-order model supported by an incomplete set of data and analyses. The possibility remains for subtle, nonlinear, biochemical controls that operate at lower levels and that are the keys to whether or not there is a trend of increasing oxygen depletion in the Bay.

To confirm or eliminate such a trend, physical influences must be quantitatively described to an accuracy that will allow a similarly accurate description of the nonphysical controls. Conceptually, once this description is complete, then the various influences can be separated. This separation process, however, is not expected to be a simple, linear extraction of component influences from time-series records. The response of Chesapeake Bay to physical forcing is a richly nonlinear process. Linear separations have proved useful as tools to gain initial insights into the underlying physics, though more complete analysis must fully incorporate nonlinear interactions. The dependence of biological and chemical processes on physical forcing is expected to be at least as nonlinear as the physics alone. When biological and chemical interactions are included, the complexity of the system will be daunting.

We have seen that scientists have developed a substantial body of knowledge on the movements of Chesapeake Bay waters and the controlling physics. While this knowledge provides some satisfaction, it is insufficient to meet the challenge that is necessary for restoration and conservation of the Bay's resources. How do we improve this level of knowledge? What should be the priority research efforts? The complexity of the problem requires both an interdisciplinary approach and the quantitative integrating power of a numerical model. Our primary research needs lie at the extreme ends of a research spectrum:

1. Increased detail and accuracy in describing small-scale processes.

2. Increased ability and accuracy in combining these descriptions of the manifold processes into a complex, interacting whole.

The dilemma facing researchers is that greater detail and complexity in the individual components of the system will make the task of building a framework to combine and unify these building blocks extraordinarily complex. Faced with such complexity, it would be too easy to view Chesapeake Bay as a chaotic, hopelessly interconnected, and ultimately unmanageable system. However, a powerful array of new observational tools such as acoustic sonars and satellite sensors, and advanced numerical and analytical techniques hold promise for achieving a predictive understanding of the flow and mixing of Bay waters and their influence on dissolved oxygen. Armed with these tools, and with the new insights that they help stimulate, scientists can address both the individual Bay processes and look towards their integration into a working whole.

Yet an increased capability will not alone suffice to solve the problem at hand. Furthermore, the present level of scientific effort is unlikely to produce the degree of understanding essential to restoration of Bay health. The primary research need to support the proposed nutrient reduction efforts has less to do with scientific capability or scientific endeavors than with the organization of the scientific effort.

A coordinated, targeted research effort is necessary for achieving a predictive understanding of the Bay's ecosystem in time to provide useful information for the policy making and resource management process. Predictive understanding refers to both the short term (hours to days) for aiding the response to environmental events and accidental spills of contaminants, and the long term for predicting conditions that result from management actions such as reduction in nutrient loading. The joint effort between the states of Maryland and Virginia is just such an example to further our understanding of these processes.

Achieving predictability, or even trend detectability, in the Chesapeake Bay ecosystem will require a quantitative understanding that is at the state of the art of present observational and analytical techniques. The incorporation of numerical models is essential for combining the myriad, coupled processes that interact in complex, nonlinear ways. For dissolved oxygen, which has been chosen as an indicator of the health of the Bay, the following specific efforts should be undertaken

1. Reexamine the 40-year oxygen record with modeling and analysis to detect physical, chemical and biological trends.

2. Supplement the future record with high-frequency monitoring.

3. Incorporate numerical modeling within the analysis of interdisciplinary process studies.

4. Establish competitions to stimulate model development.

5. Establish a specific observation-model comparison activity for model skill assessment.

6. Construct a trend-evaluation strategy to guide management.

Although the existing record is undersampled in time and contains noise resulting from short-term fluctuations in summer oxygen depletion, the provocative results from Seliger and Boggs (1988) indicate that further analysis is warranted. Their high correlation between spring runoff and oxygen-depletion volume is a challenge to reanalyze the record in an attempt to extract the biologically mediated signals by accurately removing those due to physical forcings. While the present monitoring survey frequency of 20 per year is a substantial improvement over the previous sampling strategy, recent research has shown that much of the observed change in the ecosystem does not occur in a smooth manner, but in highly episodic pulses, primarily forced by the atmosphere. Unless the present monitoring scheme is supplemented with continuous high frequency sampling from moored platforms, the ability to improve our understanding of the processes of change will be severely hampered.

The wide range of time scales of the oxygen-depletion problem and the high degree of coupling among physical and biological processes presents a daunting complexity to both observational and analytical efforts. From meteorologically driven reaeration events, to seasonal stratification and oxygen depletion, to one-year benthic nutrient retention, to decadal trends that are possibly caused by human beings — innovative and interdisciplinary techniques will be necessary to encompass and separate such a range. The integrating power of numerical modelling will be essential for the analytical phase of studies, but will also prove invaluable in the design phase of field programs.

Given the importance of modeling and analysis to the nutrient-reduction strategy of the Bay-restoration effort, the number of semi-independent modeling teams assigned to the task is inadequate. Modeling competitions such as were conducted recently for North Sea circulation are beneficial for enhancing development and improving accuracy of numerical techniques. With computer resources no longer the primary limiting factor in the costs of numerical modeling activities, the ability to assemble and target skilled labor for the task becomes increasingly important. Evaluation of the accuracy of the resulting models has traditionally been relegated to a necessary, but far insufficient, comparison of too little model output with an inadequate amount of data. We are learning that model-observation comparison and model skill assessment requires high levels of skill on the part of examiners. If modeling is set up as a major element in the decision making process for management of the Chesapeake Bay's resources, then a distinct effort in this area should be supported. Finally, an identified trend-assessment activity should be instituted. Unless we develop unambiguous criteria for evaluation of the Bay's responses to management actions, the flexibility to adjust these actions to meet the specified goals will be lost in controversy.

## Acknowledgments

This research was supported by the National Science Foundation under grant No. BSR-8814272. Carole Moore provided support on many levels, ranging from data acquisition to computer programming. Dean Pendleton helped with figure drafting. Rich Batiuk and Marsha Olson kindly provided the EPA Mainstem Monitoring data.

## References

Beardsley, R.C. and W.C. Boicourt. 1981. On estuarine and continental shelf circulation in the Middle Atlantic Bight, pp. 198-233. In: B.A. Warren and C. Wunsch (eds.) Evolution of Physical Oceanography. MIT Press, Boston.

Boicourt, W.C. 1982. The detection and analysis of the lateral circulation in the Potomac River Estuary. 1982. Publication 66, Maryland Power Plant Siting Program.

Boicourt, W.C., S.-Y. Chao, H.W. Ducklow, P.M. Glibert, T.C. Malone, M.R. Roman, L.P. Sanford, J.A. Fuhrman, C. Garside and R.W. Garvine. 1987. Physics and microbial ecology of a buoyant estuarine plume on the continental shelf. EOS 68:666-668.

Boicourt, W.C. and P. Olson. 1982. A hydrodynamic study of the Baltimore Harbor system: Part I, Observations of the circulation and mixing in Baltimore Harbor. Bulletin No. l, Ref. 82-10, Chesapeake Bay Institute, The Johns Hopkins University, Baltimore.

Boicourt, W.C. and L.P. Sanford. 1990. A hydrodynamic study of the Patuxent River estuary. Report to the Maryland Department of the Environment. Technical Report, Center for Environmental and Estuarine Studies, University of Maryland System, Cambridge, Maryland.

Brandt, A., C.C. Sarabun, H.H. Seliger and M.A. Tyler. 1986. The effects of the broad spectrum of physical activity on the biological processes in the Chesapeake Bay, pp. 361-384. In: J.C.J. Nihoul (ed.) Marine Interfaces Ecohydro-dynamics. Elsevier, Amsterdam and New York.

Carter, H.H., R.J. Regier, E.W. Schiemer and J.A. Michael. 1978. The summertime vertical distribution of dissolved oxygen at the Calvert Cliffs Generating Station: A physical interpretation. Special Report 60, Ref. 78-1. Chesapeake Bay Institute, The Johns Hopkins University, Baltimore.

Chao, S.-Y. and W.C. Boicourt. 1986. Onset of estuarine plumes. J. Phys. Oceanog. 16:2137-2149.

Chao, S.-Y and T. Paluszkiewicz. 1991. The hydraulics of density currents over estuarine sills. J. Geophys. Res. 96:7075-7076.

Chuang, W.-S. and W.C. Boicourt. 1989. Resonant seiche motion in the Chesapeake Bay. J. Geophys. Res. 94:2105-2110.

Copp, R.S. and W.C. Boicourt. 1988. The Rock Creek estuary study. Dames & Moore Report. Anne Arundel County Watershed Management Program.

Cronin, W.B. and D.W. Pritchard. 1975. Additional statistics on the dimensions of the Chesapeake Bay and its tributaries: Cross-section widths and segment volumes per meter depth. Special Report 42, Ref. No. 75-3. Chesapeake Bay Institute, The Johns Hopkins University, Baltimore.

Domotor, D.K, M.S. Haire, N.N. Panday and R.M. Summers. 1989. Patuxent estuary water quality assessment: Special emphasis 1983-1987. Report Number 43, Maryland Department of the Environment.

Doyle, B.E. and R.E. Wilson. 1978. Lateral dynamic balance in the Sandy Hook to Rockaway Point transect. Estuar. Coastal Mar. Sci. 6:165-174.

Dyer, K.R. 1973. Estuaries: A Physical Introduction. John Wiley & Sons, New York and London.

Dyer, K.R. 1977. Lateral circulation effects in estuaries, pp. 22-29. In: C.B. Officer (panel Chrmn.) Estuaries, Geophysics, and the Environment. Studies in Geophysics, National Academy of Sciences.

Dyer, K.R. 1982. Mixing caused by lateral internal seiching within a partially mixed estuary. Estuar. Coastal Shelf Sci. 15:443-457.

Dyer, KR. and A L. New. 1986. Intermittency in estuarine mixing, pp. 321-339. In: D.A. Wolfe (ed.) Estuarine Variability. Academic Press, Orlando, Florida.

Environmental Protection Agency. 1983. Chesapeake Bay Program: Findings and Recommendations. Annapolis, Maryland.

Elliott, A.J. 1978. Observations of the meteorologically induced circulation in the Potomac estuary. Estuar. Coastal Mar. Sci. 6:285-300.

Farmer, D.M. 1976. The influence of wind on the surface layer of a stratified inlet: Part II, Analysis. J. Phys. Oceanog. 6:941-952.

Farmer, D.M. and L. Armi. 1986. Maximal two-layer exchange over a sill and through the bination of a sill and contraction with barotropic flow. J. Fluid Mech. 164:53-76.

Farmer, D.M. and H. Freeland. 1983. The physical oceanography of fjords. Progr. Oceanog. 12:147-220.

Farmer, D.M. and T.R. Osborn. 1976. The influence of wind on the surface layer of a stratified inlet: Part I, Observations. J. Phys. Oceanog. 6:931-940.

Farmer, D.M. and J.D. Smith. 1978. Nonlinear internal waves in a fjord, pp. 465-493. In: J.C.J. Nihoul (ed.) Hydrodynamics of Estuaries and Fjords. Elsevier, Amsterdam and New York.

Farmer, D.M. and J.D. Smith. 1980. Tidal interaction of stratified flow with a sill in Knight Inlet. Deep-Sea Research 27:239-254.

Flemer, D.A., G.B. Mackiernan, W. Nehlsen, V.K. Tippie, R.B. Biggs, D. Blaylock, N.H. Burger, L.C. Davidson, D. Haberman, K.S. Price and J.L. Taft. 1983. Chesapeake Bay: A profile of environmental change. U.S. Chesapeake Bay Program Report, Environmental Protection Agency.

Garvine, R.A. 1977. River plumes and estuary fronts, pp. 30-35, 44. In: C.B. Officer (Panel Chrmn.) Estuaries, Geophysics, and the Environment. National Academy of Sciences, Washington, D.C.

Goodrich, D.M., W.C. Boicourt, P. Hamilton and D.W. Pritchard. 1987. Wind-induced destratification in the Chesapeake Bay. J. Phys. Oceanog. 17:2232-2240.

Haas, L.W. 1977. The effect of the spring-neap tidal cycle on the vertical salinity structure of the James, York, and Rappahannock Rivers, Virginia, U.S.A. Estuar. Coastal Shelf Sci. 5:485-496.

Heinle, D.R., C. D'Elia, J.L. Taft, J.S. Wilson, M. Cole-Jones, A. B. Caplins and L.E. Cronin. 1980. Historical review of water quality and climatic data from Chesapeake Bay with emphasis on effects of enrichment. Chesapeake Bay Program Report, U.S. Environmental Protection Agency, Washington, D.C.

Huzzey, L.M. and J.M. Brubaker. 1988. The formation of longitudinal fronts in a coastal plain estuary. J. Geophys. Res. 93:1329-1334.

Ianiello, J.P. 1977. Tidally induced residual currents in estuaries of constant breadth and depth. J. Mar. Res. 35:755-786.

Itsweire, E.C. and O.M. Phillips. 1987. Physical processes that control circulation and mixing in estuarine systems, pp. 57-73. In: M.P. Lynch and E.C. Krome (eds.) Perspectives on the Chesapeake Bay: Recent Advances in Estuarine Sciences. Publication No. 127, Chesapeake Research Consortium, Gloucester Point, Virginia.

Kuo, A.Y. and B.J. Neilson. 1987. Hypoxia and salinity in Virginia estuaries. Estuaries 10:277-283.

Kuo, A.Y., B.J. Byrne, J.M. Brubaker and J.H. Posenau. 1988. Vertical transport across an estuary front, pp. 93-109. In: J. Dronkers and W. van Leussen (eds.) Physical Processes in Estuaries. Springer Verlag, Berlin.

Malone, T.C., W.M. Kemp, H.W. Ducklow, W.R. Boynton, J.H. Tuttle and R.B. Jonas. 1986. Lateral variation in the production and fate of phytoplankton in a partially stratified estuary. Mar. Ecol. Prog. Ser. 32:149-160.

Officer, C.B., R.B. Biggs, J.L. Taft, L.E. Cronin, M. Tyler and W.R. Boynton. 1984. Chesapeake Bay anoxia: Origin, development, and significance. Science 223:22-45.

Olson, P. 1986. The spectrum of subtidal variability in Chesapeake Bay circulation. Estuar. Coastal Shelf Sci. 33:527-550.

Partch, E.N. and J.D. Smith. 1978. Time dependent mixing in a salt wedge estuary. Estuar. Coastal Mar. Sci. 6:3-19.

Pollard, R., P. Rhines and R. Thompson. 1973. The deepening of the wind-mixed layer. Geophys. Fluid Dynam. 3:381-404.

Pritchard, D.W. 1952. Salinity distribution and circulation in the Chesapeake Bay. J. Mar. Res. 11:106-123.

Pritchard, D.W. 1953. Distribution of oyster larvae in relation to hydrographic conditions, pp. 123-132. Proceedings of the Gulf Caribbean Fisheries Institute, 1952.

Pritchard, D.W. 1954. A study of the salt balance in a coastal plain estuary. J. Mar. Res. 13:133-144.

Pritchard, D.W. 1956. The dynamic structure of a coastal plain estuary. J. Mar. Res. 15:33-42.

Rattray, M., Jr. and D.V. Hansen. 1962. A similarity solution for circulation in an estuary. J. Mar. Res. 24:82-102.

Sanford, L.P. and W.C. Boicourt. 1990. Wind forced salt intrusion into a tributary estuary. J. Geophys. Res. 48:567-590.

Sanford, L.P., K.G. Sellner and D.L. Breitburg. 1990. Covariability of dissolved oxygen with physical processes in the summertime Chesapeake Bay. J. Mar. Res. 48:567-590.

Schubel, J.R. and D.W. Pritchard. 1986. Responses of upper Chesapeake Bay to variations in discharge of the Susquehanna River. Estuaries 9:236-249.

Seitz, R.C. 1971. Temperature and salinity distributions in vertical sections along the longitudinal axis and across the entrance of the Chesapeake Bay (April 1968 to March 1969). Graphical Summary Report No. 5. Ref. No. 71-7. Chesapeake Bay Institute, The Johns Hopkins University.

Seliger, H.H., J. Boggs and W.H. Biggley. 1985. Catastrophic anoxia in the Chesapeake Bay in 1984. Science 228:70-73.

Seliger, H.H. and J.A. Boggs. 1988. Long term pattern of anoxia in the Chesapeake Bay. In: M.P. Lynch and E.C. Krome (eds.) Understanding the Estuary: Advances in Chesapeake Bay Research. Publication 129. Ref. CBP/TRS 24/88, Chesapeake Research Consortium, Gloucester Point, Virginia.

Stewart, R.W. 1956. A note on the dynamic balance in estuarine circulation. J. Mar. Res. 16:34-39.

Taft, J.L. 1983. Trends in dissolved oxygen. In: Chesapeake Bay: A Profile of Environmental Change. Appendix 5. U.S. Environmental Protection Agency. Taft, J.L., E.O. Hartwig and R. Loftus. 1980. Seasonal oxygen depletion in Chesapeake Bay. Estuaries 3:242-247.

Tyler, M.A. 1984. Dye tracing of a subsurface chlorophyll maximum of a red tide dinoflagellate to surface frontal regions. Mar. Biol. 78:285-300.

Wang, D.-P. and A.J. Elliott. 1978. Non-tidal variability in the Chesapeake Bay and Potomac River: Evidence for non-local forcing. J. Phys. Oceanog. 8:225-232.

Wang, D.-P. 1979. Subtidal sea level variations in the Chesapeake Bay and relations to atmospheric forcing. J. Phys. Oceanog. 9:413-421.

Weisburg, R.H. 1976. The nontidal flow in the Providence River of Narragansett Bay: A stochastic approach to estuarine circulation. J. Phys. Oceanog. 6:721-734.

Weisburg, R.H., and W. Sturges. 1976. Velocity observations in the West Passage of Narragansett Bay: A partially mixed estuary. J. Phys. Oceanog. 6:345-354.

Zimmerman, J.T.F. 1978. Topographic generation of residual circulation by oscillatory (tidal) currents. Geophys. Astrophys. Fluid Dynam. 11:35-47.

# Effects of Water Column Processes on Dissolved Oxygen, Nutrients, Phytoplankton and Zooplankton

*Thomas C. Malone*
Horn Point Environmental Laboratory
Center for Environmental and Estuarine Studies
University of Maryland System, Cambridge, Maryland 21613

## Introduction

Assemblages of planktonic organisms form a critical link between nutrient inputs and fish production in coastal ecosystems such as Chesapeake Bay. Global scale correlations between major nutrient inputs — nitrogen, phosphorus and silicon — phytoplankton production and fish yield are an expression of this linkage (e.g., Cushing 1982; Jones and Lee 1982; Nixon and Pilson 1983). However, such correlations are often not evident on the smaller scale of coastal ecosystems where increases in nutrient input (from both natural and anthropogenic sources) have led to declines in water quality and lower fish yields. The extent to which nutrient enrichment leads to such declines rather than to enhanced fisheries production depends in part on the kinetics of the response of phytoplankton to nutrient enrichment; it also depends on the degree of coupling between phytoplankton production and the removal of this production by grazing populations such as zooplankton. Therefore, the development of predictive models relating nutrient enrichment to water quality and living resources depends on how well the rates and pathways by which organic matter is produced and consumed can be specified in time and space.

Phytoplankton productivity responds to variable nutrient supply on different time scales depending on the degree to which the rate of nutrient assimilation is limited by nutrient concentration. The growth rate over time of planktonic populations is typically a hyperbolic function of nutrient concentration, i.e., growth rate is nutrient-limited at low concentrations, but becomes nutrient-saturated at high concentrations. An important feature of this response is that nutrient-deficient cells are able to increase their rate of nutrient uptake without increasing in number. Uptake is rate limited. Once cells become nutrient-saturated, increases in nutrient uptake rate can only occur by increasing the number of cells. Uptake is limited by the number of cells or biomass present. Thus, *phytoplankton productivity exhibits a rapid (hours) response to variable nutrient supply when nutrient uptake is rate limited and a slow response (days) when nutrient uptake is biomass limited.*

Caperon et al. (1971) proposed that this fundamental relationship could be used to model the response of plankton food chains to nutrient enrichment and to define eutrophication in the context of trophic dynamics. Eutrophication is a process in which nutrients and productivity in an aquatic system increase, often leading to oxygen depletion. In an oligotrophic system (characterized by low nutrient loading), phytoplankton productivity is nutrient, or rate limited. Consequently, changes in nutrient supply are rapidly compensated for by adjustments in the rate of nutrient uptake. Ambient nutrient concentration remains low and phytoplankton biomass is stable and controlled by grazing. In contrast, in a eutrophic system (characterized by high nutrient loading), growth rate is nutrient saturated and changes in nutrient supply can only be compensated for by relatively slow changes in phytoplankton biomass. Under these conditions, phytoplankton growth rate is light- and temperature-dependent; biomass tends to be unstable; and accumulations of biomass are controlled by nutrient input. Thus, once the rate-compensating capacity of the phytoplankton community is exceeded, large fluctuations in biomass may occur in response to variable nutrient supply. Grazers such as copepods (with life cycles on the order of days to weeks) typically cannot respond quickly to these fluctuations in phytoplankton biomass, and a portion of the organic production remains ungrazed.

Such instability of plankton populations has important ecological consequences. For example, in aquatic ecosystems in which phytoplankton account for most organic input, temporal lags between phytoplankton production and zooplankton consumption result in accumulation of biomass, a phenomenon that characterizes nutrient-saturated, biomass-compensating systems. The resulting separation in time and space between photosynthetic production and heterotrophic decay is a necessary precondition for oxygen depletion (see Officer and Ryther 1977). As discussed by Verity (1987) and others (see Greve and Parsons 1977; Bird and Kalff 1984; Tuttle et al. 1987), such accumulations of ungrazed organic material are often associated with a shift from metazoan food webs that lead to fish production (e.g., phytoplankton →copepods →fishes) to microbial-dominated food webs that lead to greater decomposition and oxygen consumption (e.g., phytoplankton →bacteria →protozoa). That is, once phytoplankton production exceeds the grazing capacity of zooplankton (i.e., grazers become food saturated) the resulting accumulation of excess phytoplankton biomass sets the stage for enhanced microbial decomposition and oxygen depletion (Jonas, this volume).

Marine and estuarine systems appear to be more responsive to climatic variability than are terrestrial systems (Steele 1985). As a temperate zone estuary, Chesapeake Bay is subject to a variety of climatic forcings which influence the growth and distribution of phytoplankton and zooplankton populations as well as the distribution of dissolved oxygen. These include annual and seasonal changes in incident radiation, freshwater runoff, winds and tides. Freshwater runoff is of central importance because of its influence on nutrient input, sediment load, buoyancy reservoir, and the residence time of water and organic and inorganic matter in the Bay (see Boicourt, this volume). In this context, the mesohaline reach of Chesapeake Bay (the region between 5 to 20 ppt salinity) is of particular interest for a variety of reasons: (1) the Susquehanna River accounts for most external nutrient input (2) much of which is assimilated within this region where (3) phytoplankton production is high and (4) oxygen depletion is most severe.

As is typical of mid-latitude rivers, the annual cycle of freshwater discharge exhibits a spring maximum and a summer minimum. Consequently, 50 to 60% of the annual nutrient input to Chesapeake Bay occurs during spring runoff (March through May). Nitrate ($NO_3^-$) and particulate phosphorus are the predominant forms of nitrogen and phosphorus in freshwater discharged into the Bay. As a percentage of total external (riverine) inputs, the Susquehanna River accounts for approximately 70% of total nitrogen, 60% of total phosphorus and 60% of dissolved inorganic silicon to the Bay (Flemer et al. 1985; D'Elia et al. 1983), and about 80% and 65% of the external inputs of nitrogen and phosphorus to the mesohaline reach (reviewed by Schubel and Pritchard 1986). These rough estimates will be refined as more data become available. In particular, the data for silicon are not available for partitioning among inorganic and organic pools, although D'Elia et al. (1983) estimate that the biogenic silicon input is an order of magnitude less than the input of silicon as silicic acid. However, it is likely that the proportion of biogenic silicon varies seasonally and with river flow as reported by Anderson (1986) for the James, York and Rappahannock rivers.

Although annual phytoplankton production is apparently responsive to external nitrogen input, recent reviews (Flemer et al. 1985; Nixon 1981) indicate that these inputs support only a small fraction (10 to 20%) of phytoplankton production, the remainder being dependent on nutrient cycling within the estuary (see Kemp and Boynton, this volume). Water circulation plays a key role in the nutrient cycles of partially stratified estuaries by increasing the retention and recycling of nutrients (Redfield 1955; Ketchum 1967; Taft et al. 1978; Malone et al. 1988). The high productivity and temporal and spatial lags between nutrient inputs and phytoplankton productivity that characterize temperate estuaries have been attributed to such an interaction between circulation and nutrient cycling.

The response of phytoplankton to nutrient enrichment will also depend on the form in which nutrients are input to the system. For example, since most phosphorus enters the upper Bay bound to particles, the kinetics of phosphate desorption and regeneration will influence the time-dependent relationship between input and phytoplankton production. Lags between biogenic silicon input and subsequent phytoplankton utilization can also be expected depending on decomposition and regeneration rates. Because river input is the major source of new nutrients to the Bay and because estuarine circulation tends to retain and transport particle-bound nutrients upstream in bottom water, the effects of nutrient depletion on phytoplankton productivity are likely to be most pronounced in higher salinity water downstream of the mesohaline reach of the Bay.

This chapter focuses on the response of phytoplankton to nutrient enrichment and on the mechanisms by which plankton populations influence the relationship between nutrient enrichment and oxygen depletion in the mesohaline Chesapeake, between the Bay Bridge and the Patuxent River, where the problem is most severe. I (1) describe how distributions of phytoplankton biomass, as indicated by the concentration of chlorophyll *a*, and productivity change in time and space, (2) relate these changes to external forcings (nutrient input, incident solar radiation, wind, mixing) and internal patterns of zooplankton grazing and nutrient cycling, and (3) discuss how nutrient-phytoplankton-zooplankton interactions are related to seasonal oxygen depletion of bottom water.

## Distribution of Phytoplankton in the Chesapeake Bay

### *Biomass*

The concentration of chlorophyll *a* (Chl) is used here as an index of phytoplankton biomass. Given the goal of relating nutrient input to variations in phytoplankton biomass, the chlorophyll *a* content of the water column (mg $m^{-2}$) and of the mesohaline reach will be emphasized.

Mean monthly chlorophyll content of the mesohaline reach of Chesapeake Bay exhibits an annual cycle characterized by a spring maximum in April to May (Figure 1). Water column chlorophyll *a* levels during this period often exceed 1000 mg $m^{-2}$ over the main channel with maximum monthly means varying between 200 and 400 mg $m^{-2}$ depending on the year. These are exceptionally high chlorophyll *a* levels and approach the maximum amount of chlorophyll *a* that can be sustained by local photosynthetic growth *in situ* (see Takahashi and Parsons 1972; Wofsy 1983). Thus, chlorophyll *a* levels in excess of about 300 to 400 mg $m^{-2}$ reflect increases in biomass due to factors other than local growth, for example, accumulation of organic matter below the euphotic zone due to sedimentation of phytoplankton from surface layers or circulation patterns that accumulate or import biomass produced downstream of the Bay's mesohaline reach.

The spring chlorophyll *a* maximum in the mesohaline region occurs as a consequence of the accumulation of high chlorophyll *a* levels throughout the water column (Figures 2 and 3) with peak concentration in the lower halocline and in high salinity bottom water. The association of high chlorophyll *a* with subsurface layers suggests that phytoplankton biomass has been advected into the mesohaline Bay from the lower Bay with high salinity bottom water, a phenomenon that has been described for certain species of diatoms and dinoflagellates (e.g., Tyler and Seliger 1978; Seliger et al. 1981; Malone et al. 1988).

The spring bloom is terminated in late May to early June when bottom water chlorophyll *a* declines rapidly over a 1 to 2 week

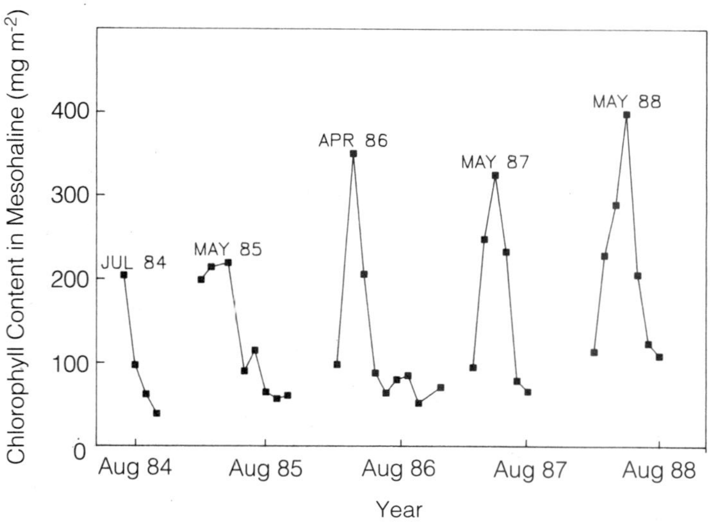


*Figure 1. Seasonal variations in the mean chlorophyll* a *(Chl) content of the water column in the mesohaline reach of Chesapeake Bay during 1984-1988; labels indicate month of maximum Chl.*

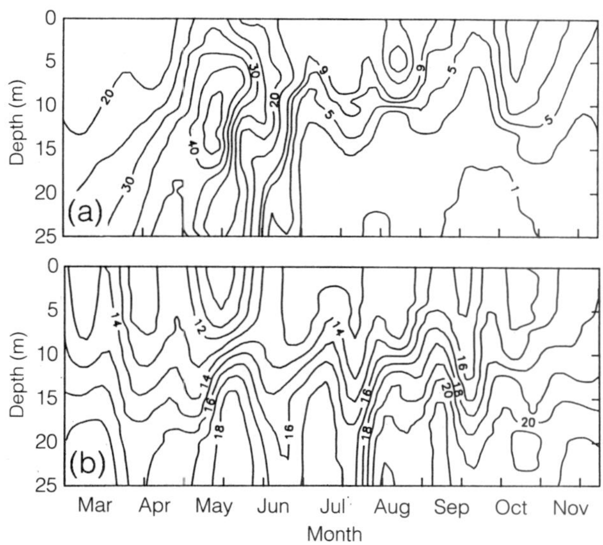


*Figure 2. Seasonal variations in the vertical distributions of (a) chlorophyll* a *(mg $L^{-1}$) and (b) salinity (ppt) at the channel station of the Chop-Pax transect in 1985.*

period. After this, high chlorophyll *a* concentrations are restricted to the surface layer throughout most of the summer (Figures 2, 4). This seasonal shift in the vertical distribution of biomass coincides with a change in the lateral distribution of chlorophyll *a*. Prior to the collapse of the spring bloom, lateral distributions of chlorophyll *a* show high concentrations throughout the water column along the Chop-Pax transect (Figure 5a, b) (see Introduction), with euphotic zone chlorophyll *a* being highest at the channel station (Table 1). After this period, chlorophyll *a* content of the euphotic zone is generally higher along the western than the eastern shore, in association with low salinity water (Figure 6). Table 1 indicates that seasonally averaged chlorophyll *a* at the westernmost station is more than double that of the easternmost station, but variability is high (Malone et al. 1986), largely as a consequence of transient blooms.

Table 1. Mean euphotic zone chlorophyll *a* content (Chl, mg $m^{-2}$) and chlorophyll specific productivity (PP/Chl, mgC $mgChl^{-1}d^{-1}$) at three stations of the Chop-Pax transect: westernmost, channel, easternmost (coefficient of variation). Based on data from 1984-1988.

| Season | Variable | West | Channel | East |
|---|---|---|---|---|
| Dec, Jan, Feb | Chl | 80 (48%) | 83 (19%) | 64 (33%) |
| | PP/Chl | 11 (91%) | 5 (56%) | 5 (57%) |
| Mar, Apr, May | Chl | 62 (47%) | 137 (57%) | 91 (71%) |
| | PP/Chl | 25 (67%) | 16 (79%) | 23 (114%) |
| Jun, Jul, Aug | Chl | 102 (105%) | 83 (64%) | 43 (70%) |
| | PP/Chl | 40 (69%) | 38 (86%) | 52 (80%) |
| Sep, Oct, Nov | Chl | 33 (55%) | 44 (45%) | 33 (64%) |
| | PP/Chl | 29 (42%) | 30 (46%) | 38 (32%) |

The transition from spring to summer chlorophyll *a* distributions coincides with a rapid change in dominant floristic groups

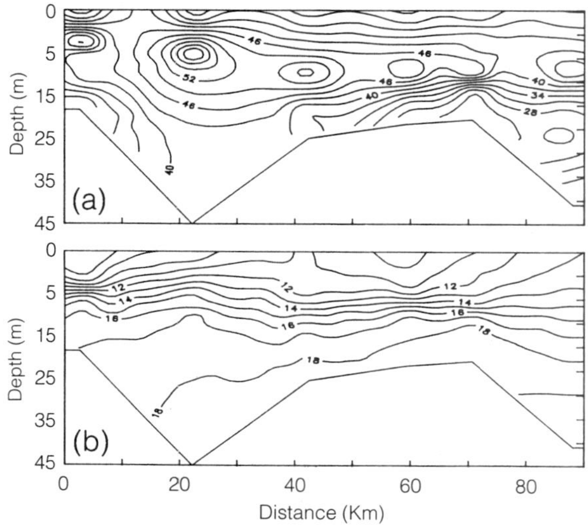


*Figure 3. Vertical distributions of (a) chlorophyll* a *and (b) salinity along a 90 km transect of channel stations (Bay Bridge to the Patuxent River) during the spring chlorophyll* a *maximum in April 1985.*

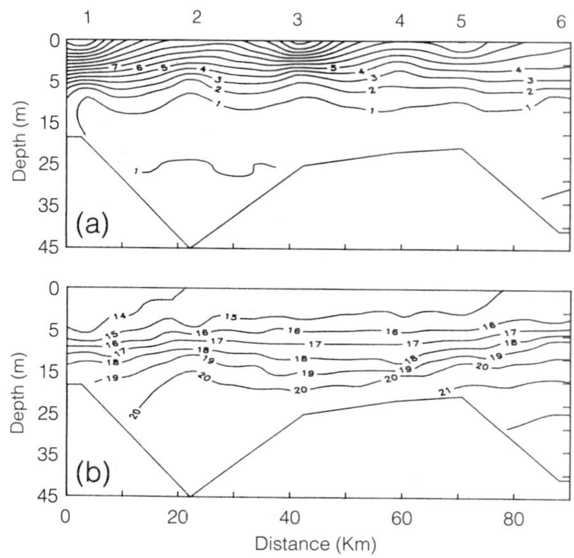


*Figure 4. Vertical distributions of (a) chlorophyll* a *and (b) salinity along a 90 km transect of channel stations (Bay Bridge to the Patuxent River) during the summer productivity maximum in August 1985.*

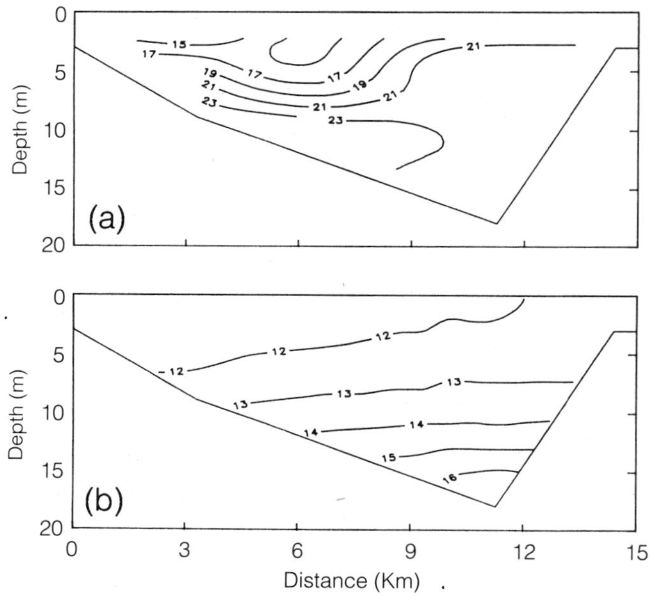


*Figure 5. Lateral distributions of (a) chlorophyll* a *and (b) salinity along the Chop-Pax transect (15 km) during spring (19 May, 1988).*

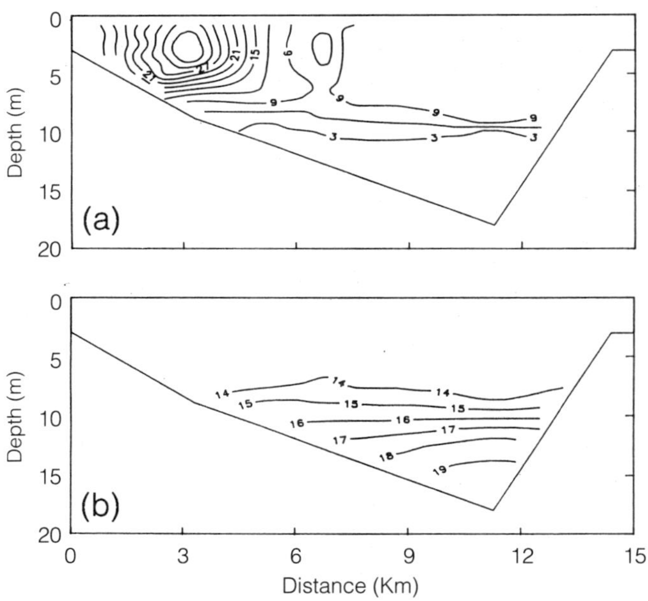


*Figure 6. Lateral distributions of (a) chlorophyll* a *and (b) salinity along the Chop-Pax transect (15 km) during summer (11 August, 1988).*

from chain-forming diatoms and large dinoflagellates to assemblages of small, solitary diatoms and cyanobacteria, chrysophytes, cryptophytes and chlorophytes as well as smaller dinoflagellates (see Van Valkenburg et al. 1978; Malone et al. 1986; Sellner 1987). Such shifts in floristic composition and patterns in the vertical and horizontal distributions of chlorophyll *a* suggest that phytoplankton distributions are dominated by diatom sedimentation, dinoflagellate motility and circulation patterns during winter-spring and by surface layer growth and grazing during summer.

### *Phytoplankton Productivity*

The annual cycle of primary productivity (PP) is characterized by a winter minimum and a summer maximum (e.g., Flemer 1970; Boynton et al. 1982; Sellner 1987). During 1984 to 1988, primary productivity in the Bay's mesohaline region usually reached its annual maximum in July or August (Figure 7a). As reported by Boynton et al. (1982), the magnitude of this primary productivity peak varies from year to year. During 1984 to 1988, mean summer primary productivity ranged from 1260 to 2560 mgC $m^{-2}d^{-1}$ (Table 2) On average, 45% of annual phytoplankton production occurred from June to August. Kemp and Boynton (1984) report that 70 to 80% of annual phytoplankton production occurred during June to September.

Table 2. Mean euphotic zone chlorophyll *a* (Chl, mg $m^{-2}$) and phytoplankton productivity (PP, mgC $m^{-2}d^{-1}$) during March-May and June-August for the Chop-Pax transect (coefficient of variation).

| Season | Variable | 1984 | 1985 | 1986 | 1987 | 1988 |
|---|---|---|---|---|---|---|
| Mar-May | Chl | – | 121 (57%) | 98(52%) | 82 (57%) | 86 (88%) |
| | PP | – | 680 (56%) | 720(48%) | 1665 (58%) | 1225 (69%) |
| Jun-Aug | Chl | 89(99%) | 52 (33%) | 71(73%) | 87 (68%) | 77 (57%) |
| | PP | 1260(80%) | 2560 (47%) | 1710(41%) | 2520 (45%) | 2030 (60%) |

The development of the summer primary productivity maximum in the mesohaline region reflects a rapid increase in chlorophyll *a* (Chl)-specific primary productivity (PP/Chl, mgC $mgChl^{-1}d^{-1}$) following the collapse of the spring bloom (Figure 7b). PP/Chl ratios increase gradually from February to May and more rapidly from May to the summer maximum. Thus, although primary productivity and chlorophyll *a* are typically correlated over the salt-intruded reach of the Bay at any given time of year (e.g., Harding et al. 1986), seasonal variations in primary productivity and chlorophyll *a* (integrated over the water column and over the mesohaline) are out of phase, with chlorophyll *a* peaking during spring and primary productivity during summer. High PP/Chl and low chlorophyll *a* relative to spring (Table 1) are indicative of the importance of grazing in controlling phytoplankton biomass during summer.

Primary productivity and PP/Chl exhibit substantial lateral variability, especially during summer (Figure 7). A detailed study of lateral variability during the summer of 1984 (Malone et al. 1986) showed that primary productivity and chlorophyll *a* were significantly correlated ($P < .01$) on the eastern and western flanks of the Bay. PP/Chl was significantly higher on the eastern than on the western flank by a factor of 3 (63 compared to 20 mgC $mgChl^{-1}d^{-1}$). A similar east-west gradient is expressed in the five-year summer mean PP/Chl (Table 1). Such opposing trends in PP/Chl and chlorophyll *a* may result from differences in grazing pressure along the eastern shore relative to the western shore (Malone et al. 1986), as well as differences in physical characteristics of water masses, species composition (K. Sellner, personal communication) and self-shading (Malone et al. 1986).

### *Picophytoplankton*

Picophytoplankton (phytoplankton passed by a 1 μm Nuclepore filter) often exhibit higher growth rates than larger phytoplankton and are too small to be effectively grazed by most metazoan consumers such as copepods. Ciliates and heterotrophic flagellates are generally thought to be the main grazers of these small phyto-

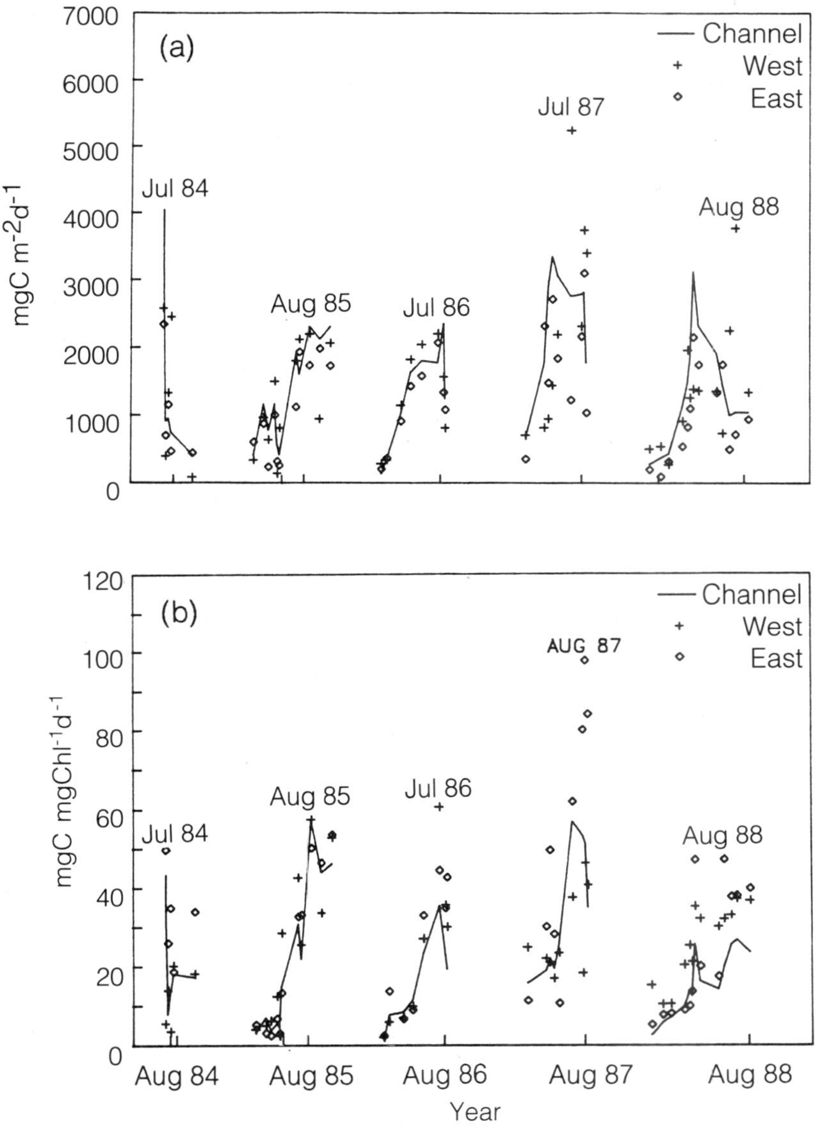


*Figure 7. Variations in (a) phytoplankton productivity (mgC $m^{-2}d^{-1}$) and (b) chlorophyll* a *specific productivity (mgC $mgChl^{-1}d^{-1}$); data from the Chop-Pax transect 1984 through 1988.*

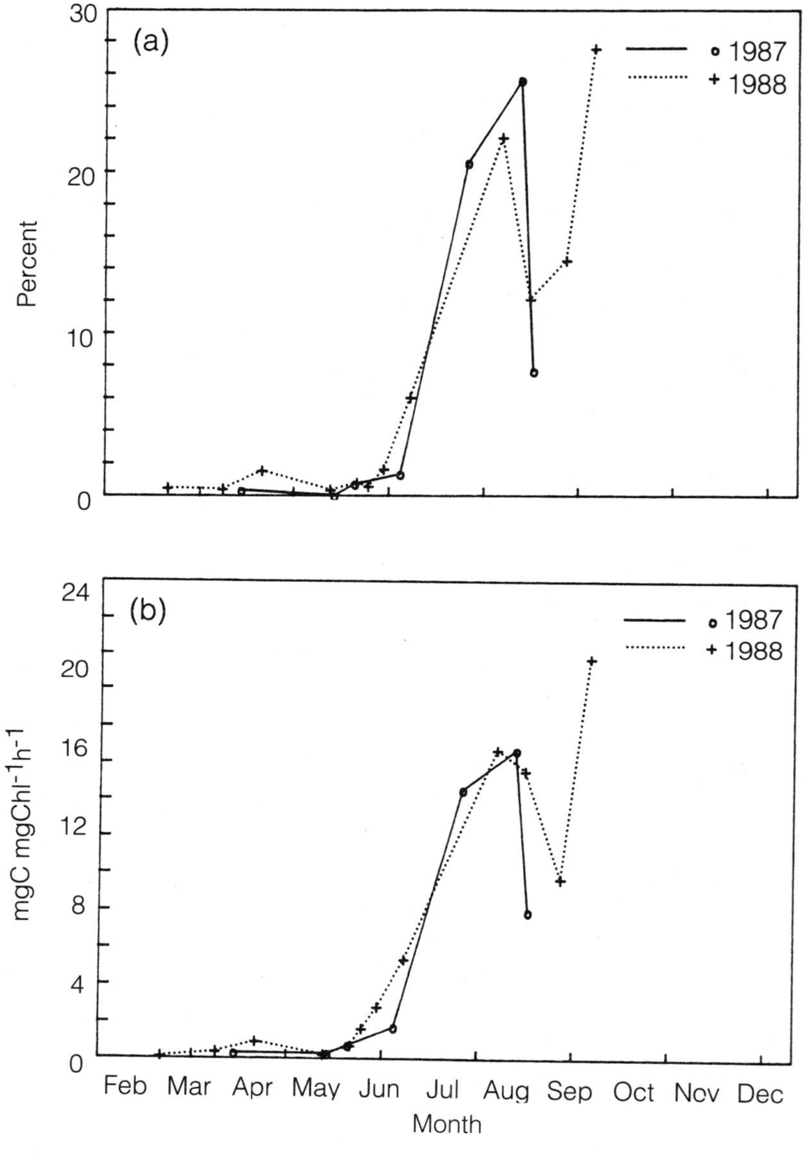


*Figure 8. (a) Percent of surface phytoplankton productivity accounted for by picophytoplankton and (b) chlorophyll* a *specific productivity of the picophytoplankton at the channel station of the Chop-Pax transect during 1987 and 1988.*

plankton (see Verity 1987; Lessard et al. 1988; Sellner 1987).

Picophytoplankton production increased from less than 2% of total phytoplankton production between February and May, to peaks in excess of 20% following the collapse of the spring bloom (Figure 8a). McCarthy et al. (1974) observed a similar annual cycle for nanophytoplankton (less than 20 µm) which accounted for over 80% of primary production and chlorophyll *a* during summer. The rapid increase in the proportion of phytoplankton productivity accounted for by picophytoplankton was associated with a marked increase in PP/Chl of picophytoplankton (Figure 8b). Thus, the transition from a phytoplankton assemblage characterized by high biomass and slow turnover during spring to an assemblage characterized by generally low biomass and more rapid and variable turnover during summer coincides with a shift in size structure with small forms becoming increasingly important.

These results provide some indication that small phytoplankton tend to have higher temperature optima than larger phytoplankton and that large diatoms may be favored by the combined effects of low temperature and high nutrient concentrations (see Goldman and Ryther 1976; Malone and Neale 1981). Alternatively, selective grazing by increased zooplankton populations on larger phytoplankton may effectively channel a greater proportion of the nutrient supply to the picophytoplankton during summer (see Steele and Frost 1977).

## Environmental Regulation of Phytoplankton

### *Nutrient Inputs and Dynamics*

Nutrient input to the mesohaline reach of the Bay is primarily driven by the flow of the Susquehanna River which typically peaks during spring (Figure 9). Given the temporal resolution of the available data, surface distributions of salinity relative to gauged flow at the Conowingo Dam (70 km upstream) indicate that the mesohaline reach responds to variations in freshwater flow with a lag of about 1 month (Malone et al. 1988; Boicourt, this volume).

Variations in the sources and sinks of inorganic nutrients required for phytoplankton growth are reflected in temporal and spatial variations in their concentration in the Bay. Using salinity as a conservative tracer of mixing between freshwater and seawater inputs to the Bay, relationships between nutrient concentration and salinity reveal important differences in the patterns of nitrate, ammonium and phosphate input to the euphotic zone.

During spring, nitrate concentration decreases rapidly with increasing salinity over a salinity range of about 5 to 20 ppt (Figure 10a). In contrast, ammonium tends to increase with increasing salinity and phosphate shows no trend (Figures 10b and 10c). During summer, nitrate concentration is generally low (less than 1 µg-at $L^{-1}$) and shows little trend with salinity while both ammonium and phosphate increase with salinity to their highest annual concentrations (Figures 11a, b and c). These distributions suggest a seasonal shift in the relative importance of the watershed and high salinity bottom water as nutrient sources to the euphotic zone, with runoff

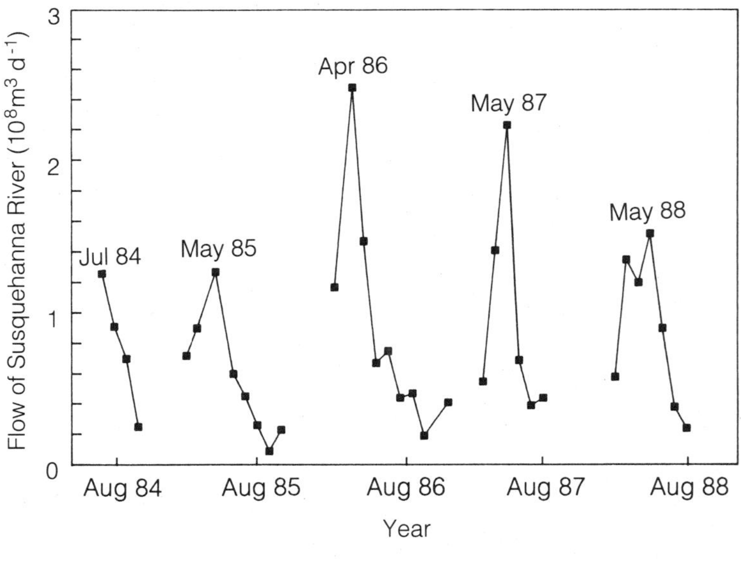


*Figure 9. Annual cycles of mean monthly flow of the Susquehanna River lagged by 1 month; labels indicate month of maximum flow.*

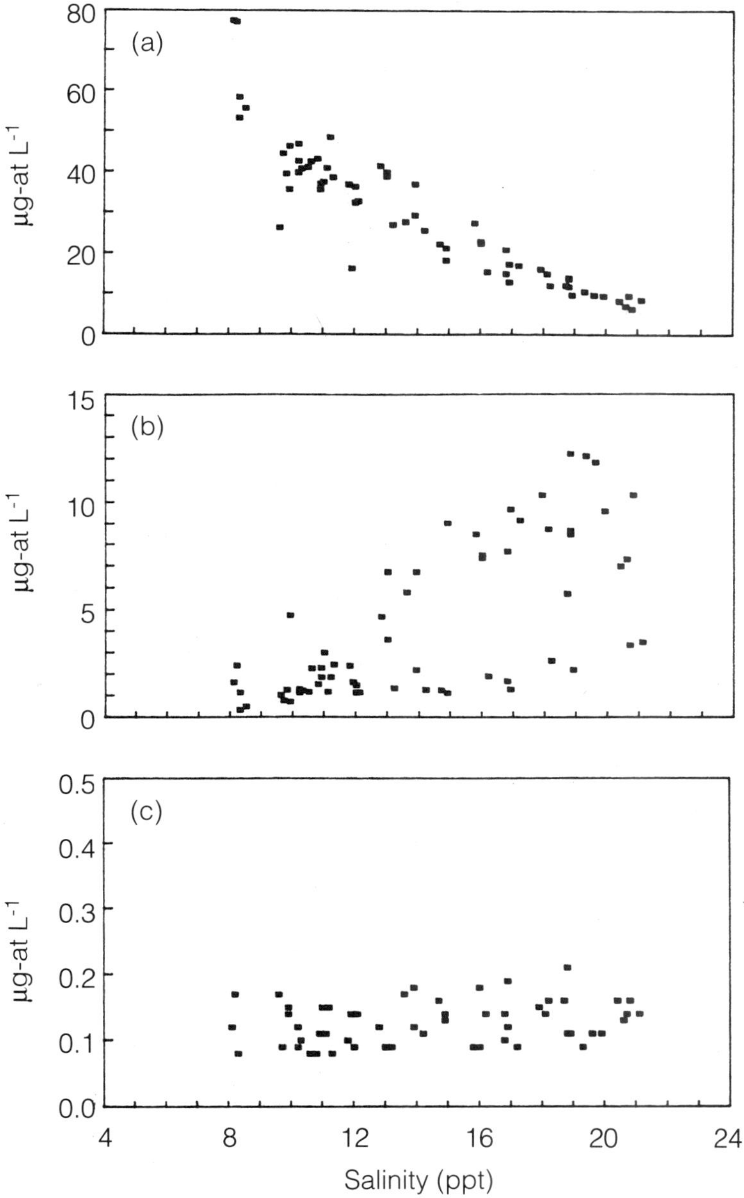


*Figure 10. Relationships between nutrient concentration (µg-at $L^{-1}$) and salinity April 16, 1986: (a) nitrate — $NO_3^-$ (b) ammonium — $NH_4^+$, and (c) phosphate — $PO_4^{3-}$.*

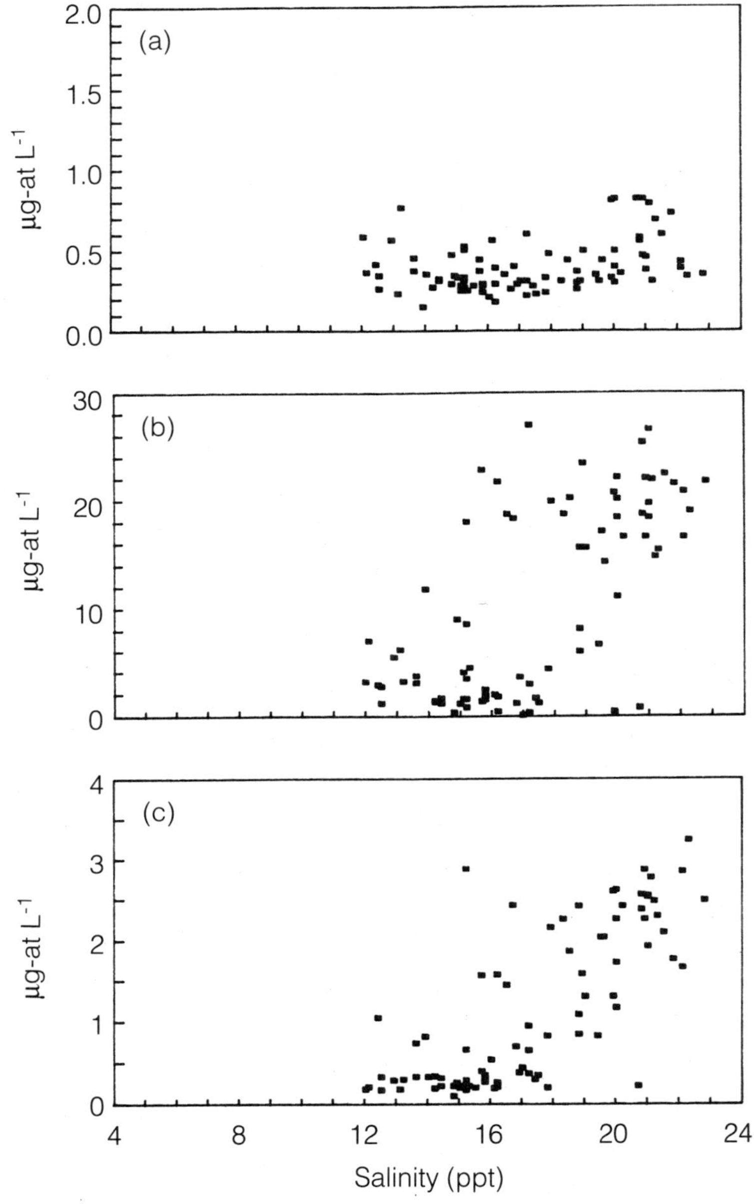


*Figure 11. Relationships between nutrient concentration and salinity August 8, 1986: (a) nitrate — $NO_3^-$ (b) ammonium — $NH_4^+$ and (c) phosphate — $PO_4^{3-}$.*

from the watershed dominating during spring and bottom water dominating during summer.

Seasonal variations in nutrient concentrations in surface and bottom water are consistent with this interpretation and indicate that high concentrations of phosphate and ammonium in bottom water are due to benthic flux (Figure 12). The annual cycle of nitrate in the surface layer is clearly driven by freshwater flow from the Susquehanna River (Figure 12a). Annual cycles of phosphate and ammonium in surface water exhibit more variability but do tend to be higher during summer than spring (Figures 10, 11, 12b and 12c). Ammonium and phosphate in bottom water exhibit pronounced annual cycles characterized by summer maxima, with ammonium increasing earlier in the year than phosphate (Figures 12b and c). Thus, nitrate input peaks during the spring as a consequence of runoff from the Susquehanna watershed while ammonium and phosphate inputs peak during the summer as a consequence of flux from the benthos, a conclusion that is consistent with results summarized by Kemp and Boynton (this volume).

It is important to realize that most of the nitrogen released as ammonium is probably derived from the original riverine input of nitrate, while much of the phosphorus released as phosphate is probably derived from the riverine input of particulate phosphorus (Kemp and Boynton 1984; Fisher et al. 1988; Malone et al. 1988). Apparently, most of the riverine input of nitrogen is assimilated by phytoplankton and transported to the benthos while much of the riverine input of phosphorus sinks to the benthos directly, without being incorporated into the biota.

As previously discussed, nitrogen input to the euphotic zone during summer is dominated by the benthic flux of ammonium. Regeneration, the major source of ammonium in the Bay, supports on the order of 80 to 90% of the annual phytoplankton production. For nutrient regeneration to have such an effect, the production and decomposition of organic matter must be separated in time and space. In Chesapeake Bay, this separation appears to occur on both seasonal and interannual time scales.

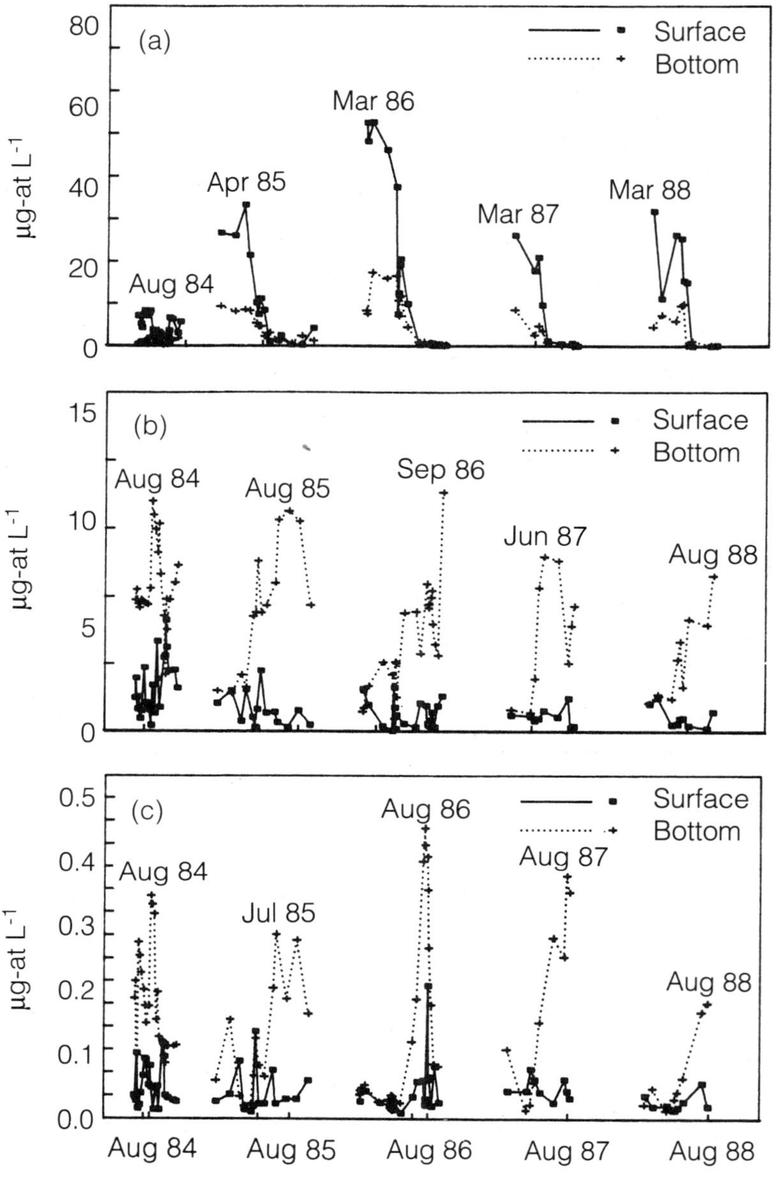


*Figure 12. Surface and bottom concentrations of (a) nitrate — $NO_3^-$ (b) ammonium — $NH_4^+$ and (c) phosphate — $PO_4^{3-}$ at the channel station of the Chop-Pax transect.*

A seasonal lag is suggested by the observation that, although 50 to 60% of the annual input of riverine nitrogen occurs during March to May, nearly 50% of annual production occurs during June to August. That is, the occurrence of maximum primary productivity during summer appears to depend on the seasonal coupling of riverine nitrate input during spring and summer ammonium regeneration. Apparently, this coupling occurs as a result of the accumulation of phytoplankton biomass during spring and the subsequent decomposition of this biomass in bottom water and the benthos (Kemp and Boynton 1984; Malone et al. 1988; Kemp and Boynton, this volume).

Conceptual models of seasonal and spatial lags between nutrient input and phytoplankton productivity have been proposed by Kemp and Boynton (1984) and Fisher et al. (1988). Spatially, nutrient uptake and phytoplankton productivity increase to maxima in the surface layer downstream of the turbidity maximum because of increased light availability. As seaward transport continues, much of the organic matter produced sinks or is transported into bottom water where upstream transport and deposition occur. Seasonally, nutrients incorporated into organic matter or adsorbed onto particles during winter-spring are retained in the estuary by a similar mechanism. As temperature increases, decomposition and regeneration result in the release of nutrients (ammonium, phosphate, silicate) from the sediments to overlying waters. The vertical transport of these nutrients then supports the development of a summer productivity maximum. Thus, although inputs of freshwater and nutrients typically peak during late winter-spring, phytoplankton productivity reaches its annual maximum during summer (Nixon 1981; Boynton et al. 1982).

### *Nutrient Limitation of Phytoplankton Production*

As a consequence of differences in the biological, chemical and physical behavior of nitrogen, phosphorus and silicon, their relative importance in limiting the production of organic matter by phytoplankton can be difficult to evaluate.[1] Nutrient concentration *per se* is a poor criterion of nutrient limitation because phytoplankton are able to attain a wide range of growth rates from near zero to rates approaching their growth maximum at nutrient concentrations that are close to the detection limits of the methods commonly used to measure nutrient concentrations. Nutrient flux is the controlling factor over the range of concentrations observed throughout most of the Bay.

Because nutrient-sufficient phytoplankton produce biomass with a nearly constant P:N:C ratio of 1:16:106 (the Redfield ratio[2]), variations in the stoichiometry of dissolved inorganic nutrients can provide clues as to which nutrient is most likely to limit production. In the Bay, nutrient ratios vary seasonally (D'Elia et al. 1983, 1986) and along the salinity gradient of the Bay (Fisher et al. 1988). N:P and Si:P ratios in freshwater entering the upper Bay (Table 3) are much higher than the Redfield ratio (N:P and Si:P of about 16:1). The annual cycle of N:P in the dissolved inorganic pool of the mesohaline Bay is characteristically higher than the Redfield ratio during the spring bloom period and lower than the Redfield ratio during the summer when productivity is highest (Figure 13). This reflects the annual cycles of nitrate input (Figure 12) and phytoplankton demand (Figure 7) and suggests that phosphorus is most likely to be depleted before nitrogen during spring but that nitrogen is most likely to be depleted first during summer. The results of nutrient enrichment experiments are consistent with this interpreta-

---

[1] The term "limiting" is used here in reference to Liebig's "Law of the Minimum" which states that the essential material (in this case the nutritional element) in the amount most closely approaching the critical minimum needed by the phytoplankton will tend to be the limiting one.

[2] The "Redfield ratio" refers to the observation by Alfred C. Redfield (1958) that atoms of phosphorus, nitrogen and carbon are, on the average, present in plankton in the ratio 1:16:106. His conclusion was based on the analysis of many samples of plankton taken from a variety of locations. The implicit assumption is that plankton, in this case phytoplankton, require phosphorus, ntirogen and carbon in this ratio for "normal" growth.

tion, leading to the hypothesis that phytoplankton are phosphorus-limited during spring and nitrogen-limited during summer (D'Elia et al. 1986; Fisher et al. 1988).

However, changes in nutrient ratios as water moves seaward through the estuary suggest that this hypothesis may not apply to the Bay as a whole. Dissolved inorganic pools of nitrogen and silicon are depleted relative to phosphorus with increasing salinity. Consequently, N:P and Si:P in water discharged from the Bay are less than biomass ratios typical of nutrient-sufficient phytoplankton (Table 3). Apparently, nitrogen and silicon are selectively lost during their transit through the estuary, presumably through denitrification (nitrogen), incorporation into refractory dissolved organic matter (nitrogen), and sedimentation (nitrogen and silicon). The comparatively conservative behavior of phosphorus has been attributed by Liss (1976) to rapid recycling and the "buffering" effect of adsorption-desorption reactions which partition phosphate between dissolved and particulate phases. These observations suggest that, on the temporal (seasonal to annual) and spatial (salt-intruded reach)

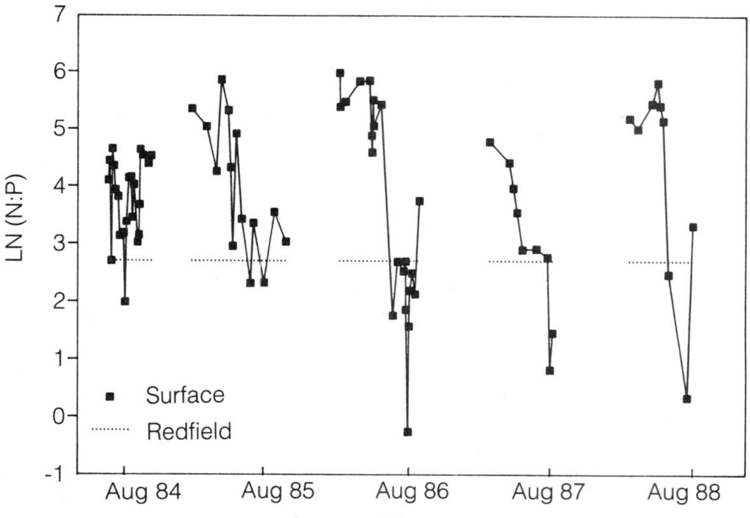


*Figure 13. Natural log transformed N:P ratio of the dissolved inorganic nutrient pool in the surface layer at the channel station of the Chop-Pax transect; note the rapid decline in summer.*

Table 3. Chemical composition of freshwater inflow of the Susquehanna River and of estuarine water outflow from the Bay and in the coastal plume of the Bay (concentrations in μg-at $L^{-1}$).

| | | | | | | Dissolved Inorganic | | | Organic |
|---|---|---|---|---|---|---|---|---|---|
| Source | Date | TN | $\%NO_3$ | TP | $\%PO_4$ | $PO_4$ | N/P | Si/P | N/P |
| Inflow[a] | Mar 82 | 166 | 70 | 1.6 | 50 | 0.8 | 163 | 105 | 44 |
| | Mar 83 | 126 | 76 | 1.6 | 18 | 0.3 | 337 | 327 | 18 |
| | Jun 82 | 127 | 92 | 1.7 | 50 | 0.9 | 133 | 101 | 8 |
| | Oct 82 | 105 | 66 | 0.8 | 9 | 0.1 | 730 | 160 | 46 |
| | Oct 83 | 119 | 30 | 1.1 | 9 | 0.1 | 410 | 320 | 78 |
| | mean | 129 | 67 | 1.4 | 27 | 0.4 | 355 | 203 | 39 |
| Outflow[a] | Mar 82 | 13 | 1 | 1.5 | 40 | 0.6 | 1.7 | 1.7 | 13 |
| | Mar 83 | 11 | 1 | 0.8 | 12 | 0.1 | 0.1 | 20.0 | 14 |
| | Jun 82 | 20 | 0 | 1.1 | 0 | 0.0 | 1.0 | 70.0 | 18 |
| | Oct 82 | 25 | 10 | 1.3 | 62 | 0.8 | 10.0 | 21.2 | 34 |
| | Oct 83 | 23 | 2 | 2.3 | 43 | 1.0 | 6.0 | 10.0 | 13 |
| | mean | 18 | 3 | 1.4 | 31 | 0.5 | 3.8 | 24.6 | 19 |
| Plume[b] | Feb 85 | – | – | – | – | 0.3 | 0.8 | 0.5 | – |
| | Apr 86 | – | – | – | – | 0.3 | 5.9 | 3.3 | – |
| | Jun 85 | – | – | – | – | 0.6 | 0.3 | 3.0 | – |
| | Aug 85 | – | – | – | – | 0.8 | 2.7 | 3.5 | – |

[a] Fisher et al. (1988)

[b] Surface concentrations for surface salinity less than 30 ppt averaged over a 2-3 week period (Garside, personal communication).

scales of the Bay, phytoplankton production is nitrogen limited, a conclusion which is consistent with the observation that annual phytoplankton production is better correlated with annual nitrogen loading than with phosphorus loading (Boynton et al. 1982).

To compound the debate over nutrient limitation in Chesapeake Bay, there is a distinct possibility that the magnitude of the spring biomass maximum and/or its rapid demise (Malone et al. 1988) may be regulated by the supply of silicate (D'Elia et al. 1983; Conley and Malone, in press). Results from the Chesapeake Bay monitoring program show that silicon is depleted relative to nitrogen and phosphorus during the development and collapse of the spring bloom (Conley and Malone, in press). The possibility of silicon limitation is of interest in the Bay because the spring bloom is dominated by diatoms, the sinking rates of which are particularly sensitive to silicon depletion (Bienfang et al. 1984). Thus, the flux of phytoplankton biomass to the benthos may be controlled by silicate input to the Bay.

## *Nutrient Regulation of Phytoplankton Biomass*

The annual cycles of river flow (Figure 9) and chlorophyll *a* (Figure 1) are remarkably similar. Water column chlorophyll *a* begins to increase in late winter as flow increases and peaks in April to May, about one month after peak discharge of the Susquehanna River. Given a one-month lag in river flow, monthly mean chlorophyll *a* content (g $m^{-2}$) of the mesohaline reach of the Bay is significantly correlated ($P < 0.01$) with monthly Susquehanna River flow, $Q_f$ ($km^3\ d^{-1}$) (Figure 14) by the least squares regression:

$$\text{(Feb-Oct)} \quad \text{chlorophyll } a = 0.03 + 1.43\ (Q_f) \quad (r^2 = 0.71,\ n = 34)$$

As the flow of the Susquehanna River accounts for most new nutrient inputs, this regression suggests that seasonal variations in phytoplankton biomass are related to the input of new nutrients, i.e., *new nutrient input is compensated for by increases in biomass rather than by increases in growth rate.* These increases in biomass lead to subsequent increases in oxygen demand.

As discussed above, biomass is more likely to be limited by nitrogen supply than by phosphorus supply, at least over the Bay as a whole. For a first-order comparison of nitrate input (Q-N) and total phytoplankton biomass in the mesohaline area, biomass as nitrogen (Ph-N) was calculated from the product of mean monthly chlorophyll *a*, N/Chl (assumed to be 10 by weight), and a surface area of $1.1 \times 10^9$ $m^2$ as described by Malone et al. (1988). Nitrate input from the Susquehanna was assumed to be the sole source of external nitrogen and was estimated as the product of river flow and nitrate concentration of the freshwater end member. The latter was estimated to be 100 μg-at $L^{-1}$ for flows greater than $10^8$ $m^3$ $d^{-1}$ and 50 μg-at $L^{-1}$ for lower flows (Smullen et al. 1982). Phytoplankton biomass as nitrogen (Ph-N) for 1984 to 1988 was significantly ($P < 0.001$) correlated with nitrate supply by the least squares regression:

$$\text{Ph-N} = 6.5 + 9.9\ (\text{Q-N}) \quad (r^2 = 0.74,\ n = 34)$$

As indicated by the slope of the regression, a unit increase in nitrate input roughly results in a 10-fold increase in phytoplankton biomass per day on average. A possible explanation for this relation is that nutrient recycling within the mesohaline and transport of biomass in bottom water from the lower Bay into the mesohaline generate the spring biomass maximum. This interpretation is consistent with the observations of Tyler and Seliger (1978) showing advective transport from the lower Bay and with results summarized by Kemp and Boynton (this volume) showing high sedimentation rates of chlorophyll *a* during the spring bloom. As discussed in the next section, much of the spring nitrogen input appears to be recycled during subsequent months, giving rise to a summer productivity maximum.

### *Regulation of Phytoplankton Productivity*

Phytoplankton productivity is influenced by a number of factors, including the concentration of chlorophyll *a*, light, temperature and nutrient supply. The similarity of annual cycles of primary productivity and incident radiation ($I_o$) suggests that seasonal variations in primary productivity are driven by incident radiation. Cole and Cloern (1984) have shown that seasonal variations in primary productivity (mgC $m^{-2}d^{-1}$) in nutrient-rich temperate estuaries are significantly correlated ($P < 0.001$) with the product of $I_o$ (E $m^{-2}d^{-1}$) and euphotic zone chlorophyll *a* (mg $m^{-2}$). The data used by Cole and Cloern for this analysis were collected from Puget Sound, the coastal plume of the Hudson River and San Francisco Bay. A similar calculation presented by Harding et al. (1986) shows that primary productivity in Chesapeake Bay is significantly correlated ($P < 0.001$) with this composite parameter by the regression:

$$PP = 254 + 0.52\ (I_o \times Chl) \quad (r^2 = 0.71,\ n = 19)$$

This regression is based on the results of experiments run during March, July and October at 5 stations from freshwater to the sea and suggests that most of the variability in primary productivity along the salinity gradient of the Bay is caused by variations in incident radiation and phytoplankton biomass.

A similar relationship describes seasonal and interannual variations in primary productivity at stations of the Chop-Pax transect (Figure 15). Based on pooled 1986 to 1988 data (the only years for which concurrent light data are available), primary productivity is significantly correlated ($P < 0.001$) with this composite parameter by the least squares regression:

$$PP = 630 + 0.31\ (I_o \times Chl) \quad (r^2 = 0.34,\ n = 87)$$

PP/Chl is also significantly correlated with incident radiation by the least squares regression:

$$PP/Chl = 7.7 + 0.46\ (I_o) \qquad (r^2 = 0.17,\ n = 87)$$

This formulation was used by Platt (1986) to show that PP/Chl is a linear function of incident radiation. Theory and empirical observations led Platt to conclude that a slope of 0.5 is characteristic of a variety of coastal and oceanic habitats. These results reflect the

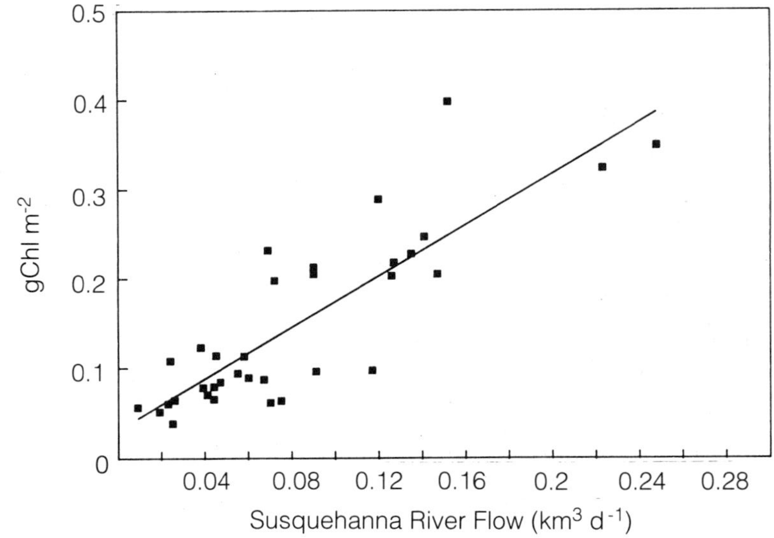


*Figure 14. Relationship between mean monthly chlorophyll* a *content of the mesohaline reach of the Bay and flow of the Susquehanna River (lagged by 1 month).*

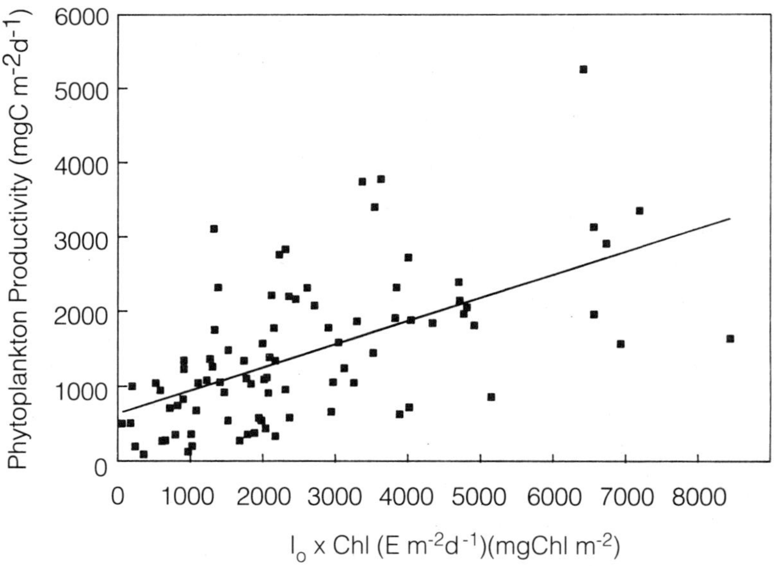


*Figure 15. Relationship between phytoplankton productivity and the composite parameter $(I_o)(Chl)$ where $I_o$ = incident photosynthetically active radiation $(E\ m^{-2}d^{-1})$ and Chl = euphotic zone chlorophyll* a *content $(mg\ m^{-2})$.*

strong influence of incident radiation and phytoplankton biomass on temporal and spatial variations in phytoplankton productivity.

However, the low $r^2$ and the large scatter in the relationship between primary productivity and the product of incident radiation and chlorophyll *a* ($I_o$ x Chl) (Figure 15) indicate that much of the variability in primary productivity is a consequence of other factors, especially during summer when incident radiation is high and productivity is most likely to be limited by water column light attenuation and/or nutrient flux and grazing. Given exponential constants for phytoplankton growth of 0.1 to 1.1 $d^{-1}$ (Table 4) and for decomposition of 0.001 to 0.1 $d^{-1}$ (Harrison 1980; Nixon and Pilson 1983), such a seasonal lag between production and decomposition is not unreasonable.

Table 4. Mean chlorophyll specific phytoplankton productivity (PP/Chl, mgC $mgChl^{-1}d^{-1}$), carbon specific growth rate ($\mu$, $d^{-1}$) and vertical density stratification ($\Delta$ $sigma_t$) with corresponding estimates of the dilution rate of the surface layer (D, $d^{-1}$) for the mesohaline reach of the Bay, slopes of regression of % $O_2$ saturation on temperature during spring (% $O_2$), and time-integrated thickness of bottom water having less than 10% saturation (meter x days) during summer; $Q_f$ = mean volume transport of the Susquehanna River, $10^8$ $m^3$ $d^{-1}$.

| Season | Year | PP/Chl | $\mu^a$ | $Q_f$ | $D^b$ | $\mu$/D | $\Delta$ $sigma_t$ | $\%O_2$ |
|---|---|---|---|---|---|---|---|---|
| Mar, Apr, May | 1984 | – | – | 2.40 | – | – | – | – |
| | 1985 | 8 | 0.16 | 1.15 | 0.08 | 2.0 | 4.8 | -4.7 |
| | 1986 | 7 | 0.14 | 1.84 | 0.10 | 1.4 | 6.6 | -4.9 |
| | 1987 | 22 | 0.44 | 1.40 | 0.09 | 4.9 | 5.0 | -8.4 |
| | 1988 | 19 | 0.38 | 1.36 | 0.09 | 4.2 | 4.6 | -4.4 |
| Jun, Jul, Aug | 1984 | 24 | 0.48 | 1.27 | 0.05 | 9.6 | 7.7 | 828 |
| | 1985 | 57 | 1.14 | 0.44 | 0.03 | 38.0 | 5.7 | 407 |
| | 1986 | 30 | 0.60 | 0.62 | 0.04 | 15.0 | 6.1 | 607 |
| | 1987 | 42 | 0.84 | 0.51 | 0.04 | 21.0 | 3.9 | 258 |
| | 1988 | 32 | 0.64 | 0.71 | 0.05 | 12.8 | 4.6 | 129 |

[a] $\mu$ = (PP/Chl) (Chl/C) where Chl/C = 0.02 by weight

[b] $D = Q_s/V$ where $Q_s = Q_f\ (S_b/[S_b-S_s])$, V = volume of surface layer ($5.24\text{x}10^9\ m^3$)

Annual phytoplankton production varies from year to year from about 250 to 600 gC m$^{-2}$, largely as a consequence of interannual variations in the magnitude of the summer productivity maximum (Boynton et al. 1982; Malone et al. 1988). It is difficult to relate such large interannual variations in summer phytoplankton productivity to differences in incident radiation, which exhibit little variability between years. Mean summer phytoplankton productivity decreases as the mean density gradient across the pycnocline increases (Figure 16). This suggests that year to year variations in summer primary productivity may be related to the vertical flux of nutrients (especially ammonium) from subpycnoclinal waters into the euphotic zone (Malone et al. 1986, 1988). Spring-neap stratification-destratification in the York River has been related to pulses of enhanced productivity following the mixing of nutrient-rich bottom waters into the euphotic zone (Haas 1977). Since summer primary productivity, which runs on recycled nutrients, accounts

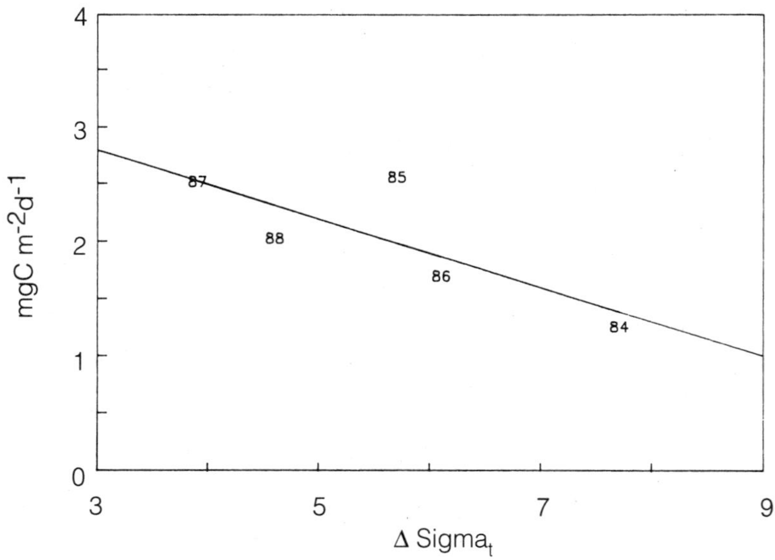


*Figure 16. Relationship between mean summer (July-August) phytoplankton productivity and vertical density stratification at the channel station of the Chop-Pax transect.*

for a larger fraction of annual primary productivity than any other season, this may explain the correlation between annual primary productivity and nitrogen-loading reported by Boynton et al. (1982).

Although seasonal variations in phytoplankton productivity do not appear to be directly coupled to river flow during any given year, indirect evidence indicates that year to year variations in productivity may also reflect longer term effects of river flow that are related to the retention and recycling of nutrients within the Bay. Phytoplankton productivity was exceptionally high during the year following high freshwater runoff caused by Tropical Storm Agnes in 1972 (Boynton et al. 1982). Using river flow as a proxy for nutrient delivery to the Bay, this suggests that nutrient input in 1972 was retained and recycled in 1973. Likewise, annual production for any given year during 1969 to 1988 is correlated with the two-year mean river flow for the respective and preceding years (W. Boynton personal communication). Thus, it appears that *nutrient retention and recycling mechanisms operate on seasonal and interannual time scales and that phytoplankton productivity is enhanced as a consequence.*

It should be emphasized that dissolved inorganic nitrogen (nitrate + nitrite + ammonium) in the surface layer is rarely depleted in the mesohaline reach of the Bay. Mean concentrations for the summer months range from 2 to 12 $\mu$g-at $L^{-1}$ with concentrations typically decreasing to about 2 $\mu$g-at $L^{-1}$ at a mean salinity of nearly 20 ppt (Fisher et al. 1988). Light levels are also more than sufficient to support phytoplankton growth in the surface layer (Harding et al. 1986). Under such conditions of light and apparent nutrient saturation, primary productivity should increase with an increase in vertical stratification rather than decrease as observed in 1984 to 1988. Thus, although interannual variability in phytoplankton production is most likely related to nitrogen loading, the mechanisms by which nutrient inputs influence phytoplankton productivity remain an elusive problem.

### *Zooplankton Grazing*

The grazing of zooplankton on particulate organic matter, including phytoplankton, is of central importance to understanding the relationship between eutrophication and fisheries. Zooplankton (1) control phytoplankton productivity by grazing and thus removing phytoplankton biomass, (2) enhance phytoplankton productivity by recycling nutrients required for phytoplankton growth, (3) influence the vertical flux of particulate organic matter by limiting the accumulation of particulate organic matter in the surface layer and by producing fecal pellets, and (4) form an important trophic link between phytoplankton and fishes. The grazing activity of zooplankton also influences energy flow and nutrient cycling through microbial food webs, both by limiting the direct flow of organic matter from phytoplankton to bacteria and by releasing dissolved organic matter via sloppy feeding and excretion (see Tuttle et al. 1987; Verity 1987; Roman et al. 1988; Jonas, this volume). Thus, the potential influence of zooplankton — from copepods to sea nettles to the larvae of benthic invertebrates — on both water quality and fisheries is substantial.

The phasing of the annual cycles of phytoplankton biomass and productivity indicates a seasonal progression from high productivity relative to grazing losses during spring to a comparatively well balanced coupling between productivity and grazing during summer. This trend is illustrated by the relationship between chlorophyll *a* content of the water column and the ratio of phytoplankton growth and dilution rate of the euphotic zone. Assuming conservation of salt on a monthly time scale and that the ratio of primary productivity to chlorophyll *a* (PP/Chl) is a reasonable index of phytoplankton growth rate (see Malone et al. 1988), phytoplankton growth rate exceeds the dilution rate of the surface layer, particularly during summer (Table 4). Thus, growth within the region has the potential of generating large and rapid increases in biomass.

If chlorophyll *a* were a simple function of the balance between growth ($\mu$) and dilution (D), chlorophyll *a* would increase as $\mu/D$ increases above unity, as observed in the Hudson Estuary (Malone 1982). However, chlorophyll *a* is inversely related to $\mu/D$ by the least squares regression:

$$\text{Chl} = 298 - 52\,(\mu/D) \qquad (r^2 = 0.58,\ n = 9)$$

Such a rapid decline in chlorophyll *a* as μ/D increases strongly implicates grazing as the major process controlling phytoplankton biomass during summer.

Despite the central role of zooplankton, the ecological role and trophic dynamics of many groups are not well understood. This is particularly true of the protozoa (ciliates and heterotrophic flagellates), the developmental stages of copepods and other invertebrates, and the gelatinous zooplankton. Results of various zooplankton grazing studies are summarized in Table 5. Macrozooplankton, dominated by copepods, apparently consume a small fraction of daily phytoplankton production during spring and up to nearly 30% of phytoplankton production during summer. The microzooplankton, composed predominantly of heterotrophic flagellates, ciliates and invertebrate larval stages, also tend to consume an increasingly large fraction of phytoplankton production as temperature increases and appear to consume a larger fraction of phytoplankton production than the macrozooplankton during all seasons.

Table 5. Percent of phytoplankton production grazed by zooplankton size classes (from Heinle 1974; Sellner 1987; White and Roman 1988; Lessard et al. 1988).

| Size Class | March | May | August |
|---|---|---|---|
| >200 mm | 2% | 3- 4% | 10- 28% |
| <200 mm | 21-42% | 13-55% | 21-153% |
| Total | 23-44% | 16-59% | 31-161% |

These trends have important implications with respect to the vertical flux of phytoplankton biomass. The seasonal increase in microzooplankton grazing parallels an increase in the importance and production of pico- and nano-phytoplankton. Since the macrozooplankton preferentially graze the larger phytoplankton, low grazing rates by this group during spring imply that most of the diatom (and perhaps the dinoflagellate) production is not consumed by higher trophic levels and is lost to the benthos.

Pigment ratios provide the most comprehensive data set on zooplankton grazing in terms of temporal and spatial resolution. Phaeophytin *a* and phaeophorbide *a* are degradation products of chlorophyll *a* excreted by zooplankton which have been grazing on phytoplankton. Consequently, the ratio of phaeopigments to chlorophyll (Phaeo/Chl) is often used as an index of grazing pressure. However, this ratio is difficult to quantitatively relate to grazing rates because of differences in temperature- and light-dependent half-lives of the degradation products. With these qualifications, the annual cycle of Phaeo/Chl appears to exhibit two peaks, the first in late May and the second in mid August (Figure 17). The August

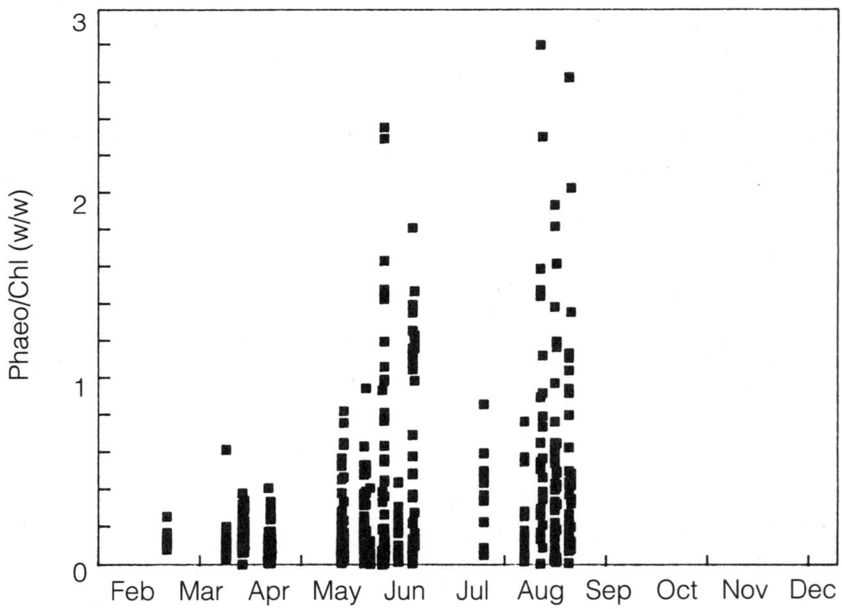


*Figure 17. The ratio phaeophorbide* a *+ phaeophytin* a*/chlorophyll* a *versus time of year along the Chop-Pax transect; pooled data 1987-1988.*

peak is consistent with measured rates of zooplankton grazing (Table 5). The May peak coincides with the collapse of the spring bloom, and measured rates of zooplankton grazing are low at this time. Thus, either the relatively few measurements of grazing rate from May are not representative or the high Phaeo/Chl peak is not related to grazing at this time but to some other process. Since phaeophytin is a natural degradation product of chlorophyll, the natural senescence and breakdown of the spring phytoplankton maximum, without involving zooplankton grazing, could cause the May phaeophytin peak.

Gelatinous zooplankton can be voracious predators. The ctenophore *Mnemiopsis* and the sea nettle *Chrysaora* are particularly important in Chesapeake Bay. *Mnemiopsis* selectively preys on copepods while *Chrysaora* preys on a variety of organisms including microzooplankton, copepods and ctenophores (see Feigenbaum and Kelly 1984; Verity 1987). Although both groups are present during the spring phytoplankton bloom, abundance maxima occur during summer when ctenophores are abundant in the main channel but are scarce on the flanks where sea nettles are extremely abundant (Table 6). This observation implies high sea nettle production in the subestuaries and their predation on ctenophores (Purcell et al. 1988).

Such alternating patterns of abundance are probably associated with major changes in the community structure of plankton communities. Feigenbaum and Kelly (1984) found that increases in the abundance of *Mnemiopsis* and *Aurelia* coincided with decreases in copepod abundance and increases in phytoplankton biomass. Increases in *Chrysaora* were associated with decreases in *Mnemiopsis* and *Aurelia*, increases in copepods, and decreases in phytoplankton biomass. Thus, predation by higher order zooplankton consumers can significantly influence phytoplankton biomass by controlling the abundance of grazing populations of copepods. However, recent results indicate that predation by gelatinous zooplankton is not a major source of mortality for copepods in the Bay (Purcell et al. 1988).

Table 6. Biomass (as determined by displacement volume, ml $m^{-3}$) of gelatinous zooplankton along the Chop-Pax transect in 1987 (Purcell et al. 1988).

| | May | | August | |
|---|---|---|---|---|
| Station | Ctenophores[a] | Hydromedusae[b] | Ctenophores[a] | Sea Nettles[c] |
| 1 | 0.6 | 1.1 | 1.5 | 10.1 |
| 2 | 1.3 | 1.8 | 10.1 | 1.7 |
| 3 | 1.1 | 1.5 | 21.0 | 1.2 |
| 4 | 1.0 | 1.1 | 1.2 | 21.5 |
| 5 | 0.7 | 3.4 | 2.1 | 21.5 |

[a] Dominated by *Mnemiopsis leidyi*
[b] Dominated by *Nemopsis bachci*
[c] *Chrysaora quinquecirrha*

## Phytoplankton Production and Oxygen Depletion

Subpycnocline waters of the mesohaline reach of the Bay undergo an annual cycle of dissolved oxygen characterized by a winter maximum and a summer minimum (Taft et al. 1980; Officer et al. 1984). Annual oxygen cycles for 1985-1988 conform to this general pattern (Figure 18). Oxygen declines during February-May from a winter maximum to less than 2 mg $L^{-1}$ by late May. Bottom water oxygen remains below 20% saturation throughout most of June to August (Figure 19).

The rapid decline in oxygen following peak freshwater flow occurs as bottom water temperature increases from 10 to 16°C and coincides with the development and collapse of the spring chlorophyll *a* maximum. This decline in bottom water oxygen (percent

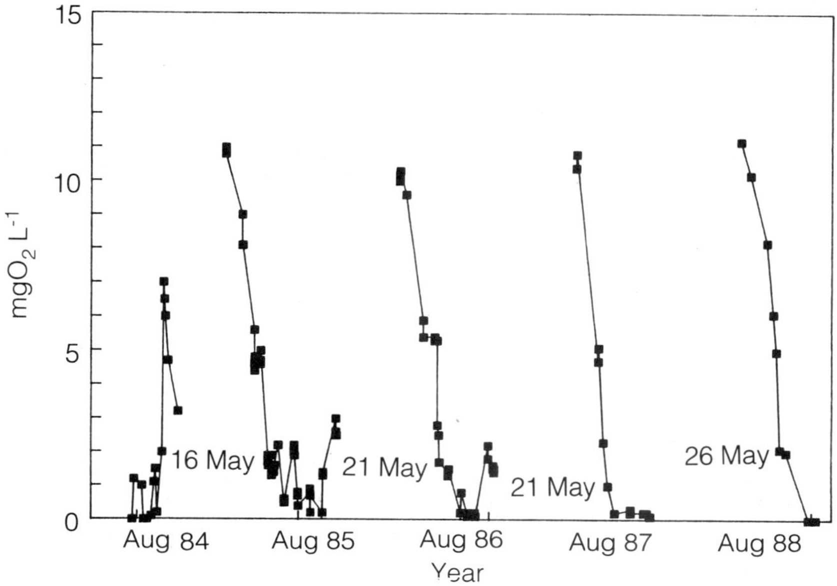


*Figure 18. Annual cycles of bottom water oxygen concentrations at the channel station of the Chop-Pax transect; labels indicate the date when dissolved oxygen fell below 2 mg L$^{-1}$.*

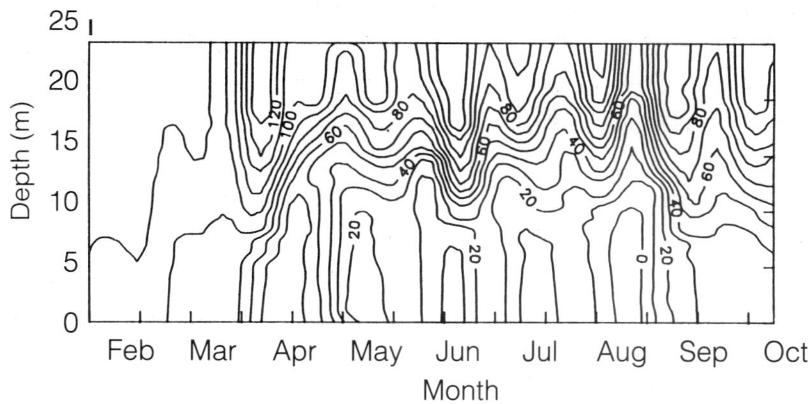


*Figure 19. Seasonal variations in the vertical distribution of percent oxygen saturation at the channel station of the Chop-Pax transect during 1985.*

saturation) during spring (March to May) is significantly correlated ($P < 0.01$) with temperature by the least squares (model II) regressions:

$$(1985)\ \%O_2 = 92.8 - 4.7\ (T) \qquad (r^2 = 0.96,\ n = 14)$$
$$(1986)\ \%O_2 = 98.0 - 4.9\ (T) \qquad (r^2 = 0.95,\ n = 17)$$
$$(1987)\ \%O_2 = 137.3 - 8.4\ (T) \qquad (r^2 = 0.84,\ n = 12)$$
$$(1988)\ \%O_2 = 108.7 - 4.4\ (T) \qquad (r^2 = 0.82,\ n = 16)$$

With the exception of 1987, the temperature-dependent rate of oxygen depletion shows little interannual variability despite large differences in vertical density stratification and freshwater flow.

The collapse of the spring bloom usually occurs as bottom water oxygen falls below 2 mg $L^{-1}$. The chlorophyll *a* content of the water column at the peak of the spring bloom varies between about 1,000-2,000 mg $m^{-2}$. Assuming stoichiometric, aerobic decay of this organic matter and a carbon to chlorophyll ratio (C/Chl) of 50 by weight (see Malone et al. 1988), these concentrations correspond to oxygen demands of about 130-260 $gO_2\ m^{-2}$, more than sufficient to account for oxygen depletion given reasonable coefficients of vertical eddy diffusivity (see Officer et al. 1984). Thus, it is not necessary to accumulate organic matter in sediments from previous years to provide the organic substrate required for seasonal oxygen depletion as suggested by Taft et al. (1980). The accumulation of phytoplankton biomass in response to nutrient input from March to May is the most likely source of organic material required to fuel the spring oxygen depletion and to maintain low oxygen levels, at least during the early stages of the summer oxygen minimum. Apparently, oxygen demand is generally not substrate limited during spring.

In contrast to the rate of oxygen decline during spring, the magnitude of summer oxygen depletion, or hypoxia, varies substantially from year to year. The volume of water having oxygen levels of less than 0.7 mg $L^{-1}$ exhibits large interannual variability (Officer et al. 1984); such variability in the areal and temporal extent of oxygen depleted bottom water appears to be related to the timing of the onset of hypoxia and on the frequency of reaeration events which occur throughout the summer (see Malone et al. 1986). As indicated by the time integrated thickness of bottom water having an oxygen concentration less than 10% of saturation (Table 4), summer oxygen depletion was most severe during the high flow year of 1984 and least severe during the drought of 1988. Variation in the extent of oxygen depletion ($O_d$, meter-days) is significantly correlated ($P < 0.05$) with mean density stratification ($\Delta$ $sigma_t$ = difference between bottom and surface densities) by the least squares (model II) regression:

$$O_d = -546 + 177 (\Delta\ sigma_t) \quad (r^2 = 0.87, n = 5)$$

Thus, as discussed by Boicourt (this volume), interannual variability in the extent of summer oxygen depletion appears to be primarily due to variations in vertical mixing and thus reaeration.

The magnitude of the spring freshet (maximum mean monthly flow in March and April, Q, $m^3\ d^{-1}$) may also influence the extent of oxygen depletion during summer as suggested by the multiple regression:

$$O_d = -546 + 129 (\Delta\ sigma_t) + 138 (Q \times 10^8) \quad (r^2 = 0.97, n = 5)$$

The influence of spring flow may reflect an increase in buoyancy of estuarine water relative to that of the adjacent coastal water, an effect that would result in greater stratification during the subsequent summer (Boicourt, this volume). Another mechanism — not mutually exclusive — involves nutrient-dependent accumulation of phytoplankton biomass between March and May and the subsequent oxygen demand caused by the decay of that biomass.

## Conclusions

The conclusions summarized below are, in my opinion, most consistent with published information. They are intended to provide a basis for constructive discussion and future research.

### *Phytoplankton and Nutrients*

Phytoplankton biomass accumulates to exceptionally high levels throughout the water column in the mesohaline reach of the Bay during spring (March to May), largely as a consequence of:

1. High nutrient input with the spring freshet.
2. Dominance of chain-forming diatoms which tend to sink from the surface layer into bottom water.
3. Two-layered estuarine flow which transports biomass into the region from the lower Bay.
4. Low zooplankton grazing rates.

The vertical flux of phytoplankton biomass to the benthos is high during this period (Kemp and Boynton, this volume).

Phytoplankton appear to be nutrient-replete as biomass accumulates during spring but may experience transient nutrient deficiency as nutrient demand exceeds supply due to the combined effects of biomass accumulation, declining river flow and nutrient uptake by freshwater phytoplankton upstream of the turbidity maximum. Such transient nutrient deficiency could develop very quickly given the large area over which biomass accumulates, the riverine source of new nutrients and high growth relative to dilution rates (Table 4). Thus, the sudden collapse of the spring bloom in late May may reflect an increase in sedimentation rates caused by the rapid development of a nutrient-deficient phytoplankton crop. Depletion of silicon may be a factor in this decline.

Zooplankton grazing appears to be an important factor limiting accumulations of phytoplankton biomass during summer. Low chlorophyll *a* concentrations in subpycnocline water indicate that direct sedimentation of phytoplankton from the surface layer is not an important loss, especially since bottom water is typically anoxic or nearly so. Predation by gelatinous zooplankton does not appear to have a major impact on copepod biomass, a finding which indicates that gelatinous zooplankton are not food-limited and do not "crop" a significant portion of the zooplankton grazers.

These observations suggest that the partitioning of phytoplankton production among microbial and metazoan food webs varies seasonally. A large fraction of the biomass produced during the spring bloom appears to be metabolized by heterotrophic microbial populations during and after the collapse of the spring bloom (Jonas, this volume). In contrast, proportionately more phytoplankton production appears to enter metazoan food chains during summer.

### *Nutrients, Phytoplankton and Dissolved Oxygen*

Coupling between nutrient input and oxygen depletion appears to occur via phytoplankton biomass and not via phytoplankton productivity *per se.* Phytoplankton biomass is directly related to seasonal variability in riverborne nutrient supply leading to large seasonal fluctuations in phytoplankton biomass, which sets the stage for seasonal oxygen depletion in the mesohaline reach of the Bay. The accumulation and sedimentation of this biomass during spring fuels spring oxygen depletion and contributes to the maintenance of summer hypoxia/anoxia. The spring biomass maximum also appears to fuel benthic ammonium regeneration which supports the development of the summer productivity maximum.

During spring, the rate of oxygen depletion in bottom water is temperature-dependent and shows little interannual variability. Apparently, the process of oxygen utilization is substrate-saturated and independent of the degree of vertical stratification over the range of conditions observed from 1985 through 1988. In contrast, the temporal and spatial extent of summer hypoxia and anoxia exhibits significant year to year variability, primarily in response to variations in vertical mixing and secondarily to variations in river flow during the previous spring, both of which are functions of climate. The magnitude of summer phytoplankton productivity does not appear to be related to the volume of oxygen-depleted water that develops during the summer.

Officer et al. (1984) presented evidence for a 30-year trend in the magnitude of summer oxygen depletion. They argued that this trend reflects an interannual increase in benthic oxygen demand caused by a nutrient-dependent increase in phytoplankton productivity. In this model, the spring decline in oxygen is driven by benthic respiration fueled by organic detritus produced during the preceding summer and fall. *Our results for 1984 to 1988 show that interannual variations in summer anoxia are dominated by climatic factors which govern freshwater input and circulation* (Boicourt, this volume). Furthermore, the effect of new nutrient input appears to be on total biomass rather than on productivity. Productivity is more responsive to nutrient retention and recycling on seasonal to annual scales than to new (riverborne) nutrient input *per se.*

In this context, the significance of silicate input as a factor limiting phytoplankton production is an important problem. The spring bloom is dominated by diatoms which account for a large fraction of the flux of organic matter to the benthos during spring. Diatoms require silica for growth and silicon deficiency causes a rapid increase in diatom sinking rates. Thus, the extent to which relative increases in nitrogen and phosphorus loading have resulted in a larger and more silicon-deficient diatom crop may be key to understanding the relationship between nutrient enrichment and oxygen depletion in the Bay.

Clearly, a nutrient management strategy should focus on reducing nutrient loading between late winter and spring. The correlation between river flow and phytoplankton biomass suggests that the rate-compensating capacity of the pelagic food web has been exceeded, resulting in large seasonal variations in biomass. The influence of climatic factors on the magnitude of summer anoxia further indicates that the aerobic capacity to assimilate phytoplankton production has also been exceeded. The extent to which such nutrient "supersaturation" of aerobic food webs has occurred should be reflected in anaerobic processes of organic decay and nutrient cycling. Thus, characterization of the rates and pathways by which phytoplankton production is metabolized is of fundamental importance to understanding the relationship between nutrient loading and oxygen depletion.

## Management Implications

The 1987 Chesapeake Bay Agreement, a landmark in the evolution of environmental management nationwide, called for a 40% reduction in nutrient loads to the Bay as a whole by the year 2000. This is an important step in the right direction, but it is not an end in itself. The success of nutrient management efforts in the watershed of the Bay and its subestuaries must be measured in terms of improved water quality and the achievement of sustainable levels of living resources. It will require several years (at least ten) of monitoring and stock assessment to establish that such trends are or are not occurring in response to management decisions.

An important conclusion of this analysis is that nonpoint nutrient inputs delivered to the Bay by the Susquehanna River support an accumulation of phytoplankton biomass that is in excess of the aerobic, oxygen assimilation capacity of the Bay under current conditions of climate and watershed use. This conclusion has several implications:

1. It is likely that reductions in nutrient loading will not be reflected in improved water quality until nutrient inputs are reduced to some as yet unknown level at which phytoplankton growth becomes rate-limited. This "unknown level" could be 10% or 90% of current inputs and will vary depending on other factors, the most important of which are climatic (e.g., precipitation, the frequency and duration of storms, mixing events), patterns of watershed use, and the response of benthic and pelagic communities to reduced nutrient loading. These unknowns underscore the importance of theoretical models, both biological and physical, as a means of establishing appropriate levels of nutrient reduction.

2. Given the sensitivity of the Bay to climatic variability, improvements in water quality, as reflected in trends such as increasing oxygen levels or sustainable fishery yields, will not

be established for several years following the implementation of a particular management practice or strategy. Thus, if management actions are to be cost-effective, they must either proceed slowly or they must increasingly be dependent on the application of empirical and theoretical models. Empirical observations of ecosystem function and development of sound ecosystem theory are thus critical to the conceptualizing of ecologically and economically sound management schemes.

3. Finally, nutrient management decisions intended to target the mesohaline reach of the Bay should be based in part on *criteria that reflect the relationships between total nutrient input (not concentration), phytoplankton biomass (not productivity),* and the temporal and spatial extent of oxygen depletion. In this context, accumulations of phytoplankton biomass required to fuel oxygen depletion appear to be more sensitive to nitrogen loading (and possibly silica inputs) than to phosphorus loading.

Clearly, limitations on point and nonpoint nutrient inputs should be a part of a broad strategy of resource management involving the conservation of water, soil and living resources. Again, the Chesapeake Bay Program is an important model for the nation. Progress is being made toward appropriate levels of sewage treatment, particularly in the subestuaries of the western shore. Increased efforts are needed to reduce nonpoint nutrient inputs, particularly during spring when precipitation and runoff from the watershed are typically highest.

## *Research Needs*

Current efforts to manage environmental changes caused by human activities tend to focus on particular attributes of ecosystems — for instance, nutrient concentrations, oxygen level, concentration of toxics and fish yield — rather than on entire ecosystems and the underlying processes that lead to undesirable changes in them, for example, species succession, patterns of energy flow and nutrient recycling. Given the need to "act now" and our inadequate understanding of these complex systems, it is important to distinguish between short-term, tactical responses to indicators of stress and long-term, strategic solutions to problems of ecosystem function. Both approaches are needed. In the short-term, limits must be placed on the rate at which we are polluting the environment. In the long-term, the development of a strategy for environmental management must be based on a sound theoretical understanding of ecosystems and the role of the human species in ecosystems.

The achievement of a truly integrated program of environmental management is a long-term goal that will require a better fundamental understanding of how the watershed and its rivers and estuaries respond to inputs of energy and nutrients in terms of the rates and pathways of nutrient cycling and export, trophic dynamics of microbial and metazoan food webs, and the relationship of these processes to water quality and fishery yields. In this context, major gaps in information include the following:

1. How changes in the patterns of watershed use influence freshwater runoff, erosion (sediment transport) and nutrient inputs (amount, form and distribution) to the Bay and its subestuaries.

2. The mechanisms by which the accumulation and turnover of phytoplankton biomass is influenced by external inputs (freshwater runoff, sediment and nutrient loading) relative to internal changes (circulation, losses to pelagic and benthic grazers).

3. The processes that govern the partitioning of phytoplankton production among microbial and metazoan food webs and how such partitioning is related to water quality and fisheries.

These are ecosystem level problems that must be addressed through coordinated interdisciplinary research and monitoring programs. As simply stated by Aristotle (Stone 1988) ". . . the doctor does not treat the disease, he treats the patient."

## Acknowledgments

The research reported here was supported by the NOAA Sea Grant Program, the U.S. Environmental Protection Agency and the National Science Foundation (LMER Program, BSR-8814272). I am grateful to M. Roman, J. White and J. Purcell for allowing me to use unpublished data. K. Sellner and L.W. Harding, Jr. provided constructive criticism. This work would not have been possible without the valuable assistance of S. Pike, S. Kelly and B. Wendler.

## References

Anderson, G.F. 1986. Silica, diatoms and a freshwater productivity maximum in Atlantic coastal plain estuaries, Chesapeake Bay. Estuar. Coast. Shelf Science 22:183-197.

Bienfang, P.K., P.J. Harrison and L.M. Quarmby. 1982. Sinking rate response to depletion of nitrate, phosphate and silicate in four marine diatoms. Mar. Biol. 67:295-302.

Bird, D.F. and J. Kalff. 1984. Empirical relationships between bacterial abundance and chlorophyll concentration in fresh and marine waters. Can. J. Fish. Aquat. Sci. 41:1015-1023.

Boynton, W.R., W.M. Kemp and C.W. Keefe. 1982. A comparative analysis of nutrients and other factors influencing estuarine phytoplankton production, pp. 69-90. In: V.S. Kennedy (ed.) Estuarine Comparisons.

Boynton, W.R. and W.M. Kemp. 1985. Nutrient regeneration and oxygen consumption by sediments along an estuarine salinity gradient. Mar. Ecol. Prog. Ser. 23:45-55.

Boynton, W.R., J. Garber, J. Barnes and W.M. Kemp. 1988. Ecosystem Processes Component. Level I Report to State of Maryland, Department of the Environment. University of Maryland Center for Environmental and Estuarine Studies, Solomons. Ref. No. (UMCEES) CBL 88-2.

Caperon, J., S.A. Cattell, and G. Krasnick. 1971. Phytoplankton kinetics in a subtropical estuary: eutrophication. Limnol. Oceanogr. 16:599-607.

Carpenter, J.H., D.W. Pritchard, and R.C. Whaley. 1969. Observations of eutrophication and nutrient cycles in some coastal plain estuaries, pp. 210-221. In: Eutrophication: Causes, Consequences and Correctives. Proceedings of Symp. Natl. Acad. Sci. Publ. 1700.

Cole, B.E. and J.E. Cloern. 1984. Significance of biomass and light availability to phytoplankton productivity in San Francisco Bay. Mar. Ecol. Prog. Ser. 17:15-24.

Conley, D.J. and T.C. Malone. In press. The annual cycle of dissolved silicate in Chesapeake Bay: Implication for production and fate of phytoplankton biomass. Mar. Ecol. Prog. Ser.

Cushing, D.H. 1982. Climate and Fisheries. Academic Press, London.

D'Elia, C.F., D.M. Nelson and W.R. Boynton. 1983. Chesapeake Bay nutrient and phytoplankton dynamics: III. The annual cycle of dissolved silicon. Geochimica Cosmochimica Acta 47:1945-1955.

D'Elia, C.F., J.G. Sanders and W.R. Boynton. 1986. Nutrient enrichment studies in a coastal plain estuary: phytoplankton growth in large-scale, continuous cultures. Can. J. Fish. Aquat. Sci. 43:397-406.

Feigenbaum, D. and M. Kelly. 1984. Changes in the lower Chesapeake Bay food chain in the presence of the sea nettle *Chrysaora quinquecirrha*. Mar. Ecol. Prog. Ser. 19:39-47.

Fisher, T.R., L.W. Harding, D.W. Stanley and L.G. Ward. 1988. Phytoplankton, nutrients, and turbidity in the Chesapeake, Delaware, and Hudson Estuaries. Estuar. Coast. Shelf Sci. 27:61-93.

Flemer, D.A. 1970. Primary production in Chesapeake Bay. Chesapeake Sci. 11: 117-129.

Flemer, D.A., W.R. Boynton, C.F. D'Elia, W.M. Kemp, M. Nichols, R.J. Orth, J.T. Smullen, J. Taft and R.L. Wetzel. 1985. The Chesapeake Bay Program: A summary of scientific research to address management needs for Chesapeake Bay, pp. 399-438. In: N.L. Chao and W. Kirby-Smith (eds.) Proc. Intern. Symp. on Utilization of Coastal Ecosystems: Planning, Pollution and Productivity, 21-27 November 1982. Rio Grande, Brazil.

Goldman, J.C. and J.H. Ryther. 1976. Temperature-influenced species competition in mass cultures of marine phytoplankton. Biotechnol. Bioeng. 18:1125-1144.

Greve, W. and T.R. Parsons. 1977. Photosynthesis and fish production: hypothetical effects of climatic change and pollution. Helgolander wiss. Meeresunters. 30:666-672.

Haas, L.W. 1977. The effect of the spring-neap tidal cycle on the vertical salinity structure of the James, York and Rappahannock Rivers, Virginia, U.S.A. Estuar. Coast. Mari. Sci. 5:458-496.

Harding, L.W., B.W. Meeson and T.R. Fisher. 1986. Phytoplankton production in two east coast estuaries: Photosynthesis-light functions and patterns of carbon assimilation in Chesapeake and Delaware Bays. Estuar. Coast. Shelf Sci. 23:773-806.

Harrison, W.G. 1980. Nutrient regeneration and primary production in the sea, pp. 433-460. In: P.G. Falkowski (ed.) Primary Productivity in the Sea. Plenum Press, New York.

Heinle, D.R. 1974. An alternate grazing hypothesis for the Patuxent Estuary. Ches. Sci. 15:146-150.

Jones, R.A. and G.F. Lee. 1982. Recent advances in assessing the impact of phosphorus loads on eutrophication-related water quality. Water Res. 16:503-515.

Kemp, W.M. and W.R. Boynton. 1981. External and internal factors regulating metabolic rates of an estuarine benthic community. Oecologia 51:19-27.

Kemp, W.M. and W.R. Boynton. 1984. Spatial and temporal coupling of nutrient inputs to estuarine production: the role of particulate transport and decomposition. Bull. Mar. Sci. 35:242-247.

Ketchum, B.H. 1967. Phytoplankton nutrients in estuaries, pp. 329-335. In: G.H. Lauff (ed.) Estuaries. AAAS Publ. No. 83.

Lessard, E.J., D.C. Brownlee and D.A. Caron. 1988. Role of microzooplankton biomass, growth and grazing in plankton dynamics in C.B. [Abstract only] EOS 69(44):1103.

Liss, P.S. 1976. Conservative and non-conservative behavior of dissolved constituents during estuarine mixing, pp. 93-130. In: J.D. Burton and P.S. Liss (eds.) Estuarine Chemistry. Academic Press, London and New York.

Magnien, R.E., K.G. Sellner and P.A. Vaas. 1988. Nutrient control of phytoplankton production in the C.B. mainstem and tributaries. [Abstract only] EOS 69(44):1103.

Malone, T.C. 1982. Phytoplankton photosynthesis and carbon-specific growth: Light-saturated rates in a nutrient-saturated environment. Limnol. Oceanogr. 27:226-235.

Malone, T.C. and P.J. Neale. 1981. Parameters of light-dependent photosynthesis for phytoplankton size fractions in temperate estuarine and coastal environments. Mar. Bio. 61:289-297.

Malone, T.C., W.M. Kemp, H.W. Ducklow, W.R. Boynton, J.H. Tuttle and R.B. Jonas. 1986. Lateral variation in the production and fate of phytoplankton in a partially stratified estuary. Mar. Ecol. Prog. Ser. 32:149-160.

Malone, T.C., L.H. Crocker, S.E. Pike and B.W. Wendler. 1988. Influences of river flow on the dynamics of phytoplankton production in a partially stratified estuary. Mar. Ecol. Prog. Ser. 48: 235-249.

McCarthy, J.J., W.R. Taylor and M.E. Loftus. 1974. Significance of nanoplankton in the Chesapeake Bay estuary and problems associated with the measurement of nanoplankton productivity. Mar. Biol. 24: 7-16.

McCarthy, J.J., W.R. Taylor and M.E. Loftus. 1977. Nitrogenous nutrition of the plankton in the Chesapeake Bay. I. Nutrient availability and phytoplankton preferences. Limnol. Oceanogr. 22:996-1011.

Nixon, S.W. 1981. Remineralization and nutrient cycling in coastal marine ecosystems, pp. 111-138. In: B.J. Neilson and L.E. Cronin (eds.) Estuaries and Nutrients. Humana Press, Clifton, New Jersey.

Nixon, S.W. and M.G. Pilson. 1983. Nitrogen in estuarine and coastal marine ecosystems, pp. 565-648. In: E.J. Carpenter and D.G. Capone (eds.) Nitrogen in the Marine Environment. Academic Press, New York.

Officer, C.B. and J.H. Ryther. 1977. Secondary sewage treatment versus ocean outfalls: an assessment. Science 197:1056-1060.

Officer, C.B., R.B. Biggs, J.L. Taft, L.E. Cronin, M. Tyler and W.R. Boynton. 1984. Chesapeake Bay anoxia: Origin, development, and significance. Science 223:22-27.

Platt, T. 1986. Primary production of the ocean water column as a function of surface light intensity: Algorithms for remote sensing. Deep-Sea Res. 33:149-163.

Purcell, J.E., M.R. Roman and J.R. White. 1988. Effects of ctenophore and scyphomedusa predators on herbivores in Chesapeake Bay [Abstract only] EOS 69(44):1103.

Redfield, A.C. 1955. The hydrology of the Gulf of Venezuela. Deep Sea Res. 2 (Suppl.):115-133.

Redfield, A.C. 1958. The biological control of chemical factors in the environment. Am. Scientist 46(3):205-221.

Roman, M.R., K.A. Ashton and A.L. Gauzens. 1988. Day/night

differences in the grazing impact of marine copepods. Hydrobiogia 167/168:21-30.

Schubel, J.R. and D.W. Pritchard. 1986. Responses of upper Chesapeake Bay to variations in discharge of the Susquehanna River. Estuaries 9:236-249.

Seliger, H.H., K.R. McKinley, W.H. Biggley, R.B. Rivkin and K.R.H. Aspden. 1981. Phytoplankton patchiness and frontal regions. Mar. Biol. 61:119-131.

Seliger, H.H., J.A. Boggs and W.H. Biggley. 1985. Catastrophic anoxia in the Chesapeake Bay in 1984. Science 228:70-73.

Sellner, K.G. 1987. Phytoplankton in the Chesapeake Bay: role in carbon, oxygen, and nutrient dynamics, pp. 134-157. In: S.K. Majumdar, L.W. Hall, and H.M. Austin (eds.) Contaminant Problems and Management of Living Chesapeake Bay Resources. Pennsylvania Academy of Science, Philadelphia.

Smullen, J.T., J.L. Taft and J. Macknis. 1982. Nutrient and sediment loads to the tidal Chesapeake Bay system, pp. 147-262. In: Chesapeake Bay Technical Studies: A Synthesis. U.S. E.P.A., Annapolis, Maryland.

Steele, J.H. 1985. A comparison of terrestrial and marine ecological systems. Nature 313:355-358.

Steele, J.H. and B.W. Frost. 1977. The structure of plankton communities. Phil. Trans. Roy. Soc. Lond. B. 280:485-534.

Stone, I.F. 1988. The Trial of Socrates. Doubleday, New York.

Taft, J.L., A.J. Eliot and W.R. Taylor. 1978. Box model analysis of Chesapeake Bay ammonium and nitrate fluxes, pp. 115-130. In: M.L. Wiley (ed.) Estuarine Interactions. Academic Press, New York.

Taft, J.L., E.O. Hartwig and R. Loftus. 1980. Seasonal oxygen depletion in Chesapeake Bay. Estuaries 3:242-247.

Takahashi, M. and T.R. Parsons. 1972. Maximization of the standing stock and primary productivity of marine phytoplankton under natural conditions. Indian J. Mar. Sci. 1:61-62.

Tuttle, J.H., R.B. Jonas and T.C. Malone. 1987. Origin, development and significance of Chesapeake Bay anoxia, pp. 442-472. In: S.K. Majumdar, L.W. Hall, and H.M. Austin (eds.) Contaminant Problems and Management of Living Chesapeake Bay Resources. Pennsylvania Academy of Science, Philadelphia.

Tyler, M.A. and H.H. Seliger. 1978. Annual subsurface transport of a red tide dinoflagellate to its bloom area: Water circulation patterns and organism distributions in the Chesapeake Bay. Limnol. Oceanogr. 23:227-246.

Van Valkenburg, S.D., J.K. Jones and D.R. Heinle. 1978. A comparison by size classes and volume of detritus versus phytoplankton in Chesapeake Bay. Estuar. Coast. Mar. Sci. 6:569-582.

Verity, P.G. 1987. Factors driving changes in the pelagic trophic structure of estuaries, with implications for the Chesapeake Bay, pp. 35-56. In: M.P. Lynch and E.C. Krome (eds.) Perspectives on the Chesapeake Bay: Advances in Estuarine Sciences, Publication No. 127. Chesapeake Research Consortium, Gloucester Point, Virginia.

White, J.R. and M.R. Roman. 1988. Grazing and egg production by Chesapeake Bay zooplankton in spring and summer, p. 78. In: M.P. Lynch and E.C. Krome (eds.) Understanding the Estuary: Advances in Chesapeake Bay Research, Publication No. 129. Chesapeake Research Consortium, Gloucester Point, Virginia.

Wofsy, S.C. 1983. A simple model to predict extinction coefficients and phytoplankton biomass in eutrophic waters. Limnol. Oceanogr. 28:1144-1155.

# Microbial Processes, Organic Matter and Oxygen Demand in the Water Column

*Robert Jonas*
Department of Biology, George Mason University
Fairfax, Virginia 22030

## Introduction

Bacteria have long been known to play an important ecological role in aquatic systems; however, except in specialized ecosystems or in cases of pathogenic or malodorous conditions, bacterial *processes* have often been ignored. In part, this reflects the public's concern with harvestable resources, which has meant that, historically, biological research focused on the more obvious factors controlling the availability of those resources. In addition, aquatic microbial ecology lacked the sophisticated methodological tools necessary for studying bacterial abundance, distribution and activity. In the last decade, though, new techniques have been developed and refined for conducting these studies (see Atlas and Bartha 1987 for a review of current methods). Such techniques have enabled researchers to frame and investigate the following hypotheses about the role of microorganisms in hypoxic and anoxic processes in Chesapeake Bay (Figure 1).

1. Heterotrophic bacteria, microbial organisms that break down organic matter, are responsible, either directly or indirectly, for most of the oxygen consumption in Chesapeake Bay.

2. Water column bacterial processes are important, compared with sediment processes, in controlling dissolved oxygen concentrations.

3. The main source of organic matter is autochthonous phytoplankton production in the Bay, though this production might be separated spatially and temporally from the oxygen consumption.

4. Both particulate and dissolved organic matter could be important contributors to biochemical oxygen demand (BOD).

To test these hypotheses, researchers asked two basic types of questions, namely, (1) How many bacteria are present, where are they, and how active are they? (2) Do bacteria consume the dissolved oxygen? In addition, they wanted to know about the organic matter which supports oxygen consumption, in particular, its physical and chemical nature and its origins; and, finally, whether or not there is an important bacterially-based food web which influences the trophic structure of Chesapeake Bay.

This chapter discusses the role bacteria play in controlling oxygen consumption in the Bay. It approaches this task by examining the various experimental approaches for measuring bacterial community characteristics, for example, abundance, production and metabolism; the ecological role of bacterioplankton (the free-living bacteria in the water column); and the relationships between bacterial processes and changes in food web dynamics.

## Experimental Approaches for Studying Aquatic Microbial Processes

Beginning with the assumption that bacteria in the water column were important to oxygen consumption, researchers had expectations of high abundances of very active bacteria there. It seemed likely they would be especially prominent near the pycnocline, where the rate of sinking of floating materials, such as detritus (particulate matter, generally of organic origin) and dead or dying phytoplankton and zooplankton, would be slowed. In addition, it seemed likely

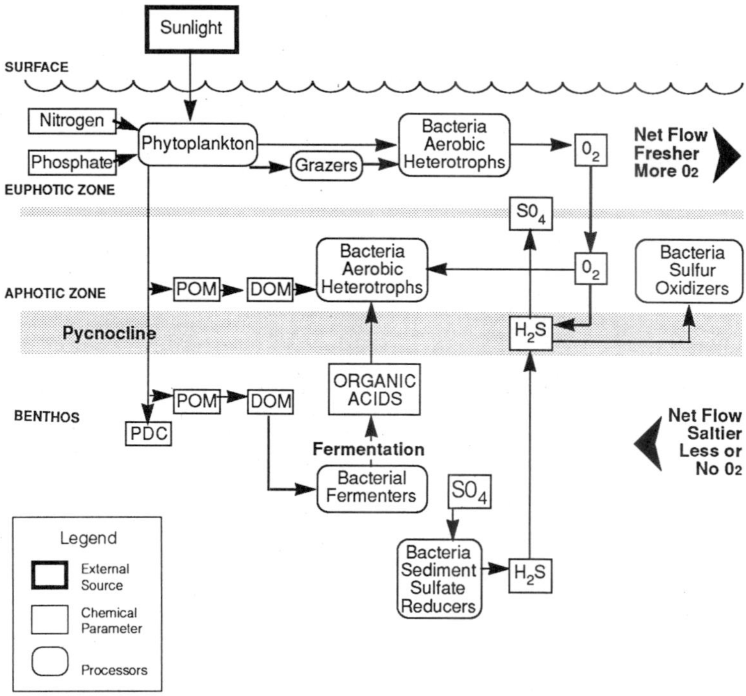


*Figure 1. Interactions among phytoplankton, bacterioplankton, sediment bacterial processes, organic matter and oxygen concentrations in the stratified mesohaline reach of Chesapeake Bay.*

that the highest abundances and activities would be in geographic areas of greatest dissolved oxygen consumption or oxygen deficit.

With regard to the sources of organic matter necessary for supporting bacterial communities: (1) if phytoplankton are the direct source, then bacterial abundance and/or activity should correlate strongly with phytoplankton variables; (2) if phytoplankton are the indirect source of organic matter, then bacterial parameters might correlate positively with chlorophyll degradation products (e.g., phaeopigments, phytodetritus) or with dissolved organic matter (e.g., amino acids or carbohydrates) in the water column.

Experimentally, a suite of measurements was taken at each station located along transects across the Chesapeake Bay mainstem (Figure 2). These stations, including the Chop-Pax transect (see Introduction), were occupied at varying time intervals ranging from a few days to weeks. Measurements included depth distributions of bacterial abundance, bacterial production and metabolism (incorporation alone or incorporation and respiration) of specific organic substrates, biochemical oxygen demand (BOD), dissolved BOD (passing a Gelman Type A/E filter, a standard filter for determination of suspended solids in water), particulate organic carbon (POC) and particulate organic nitrogen (PON) as well as phytoplankton parameters, nutrient concentrations and hydrodynamic water quality characteristics (Malone et al. 1986; Tuttle et al. 1987a, b; Jonas et al. 1988b; Ducklow et al. 1988). Tuttle et al. (1987a) investigated relevant aspects of microbial sulfur cycling at selected stations in separate efforts and in association with those noted above.

In addition, various researchers employed other experimental designs. Zieman and Macko (1988) and Lagera and Zieman (1988) investigated the role of macrophyte detritus and its decomposition, in oxygen dynamics. Kemp (1987) partitioned water column respiration into important size classes to establish the nature of the organisms contributing to the oxygen consumption process.

Based on the transect studies, it was possible to develop an east/west and north/south seasonal picture of the dynamics of the microbial and organic carbon components of oxygen demand in the Bay. Other specific hypotheses regarding processes of potential im-

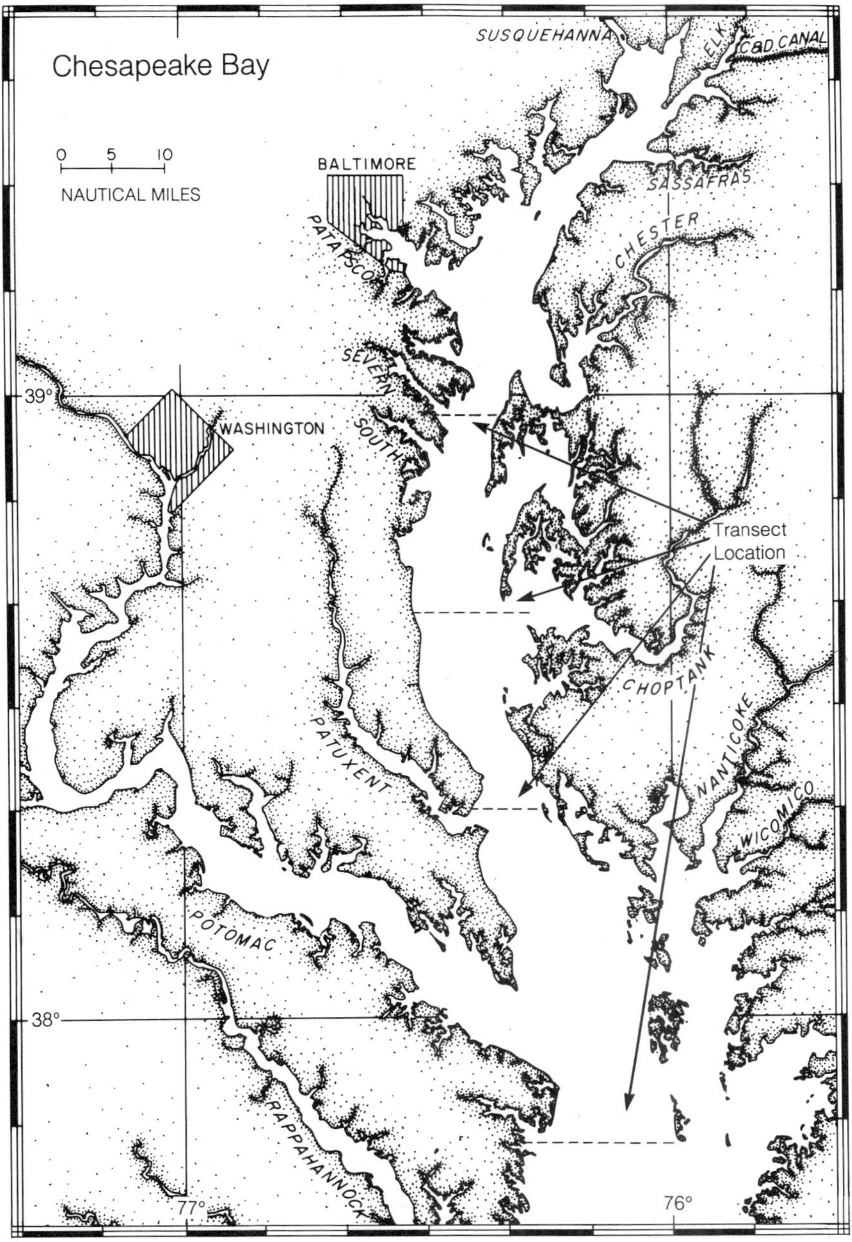


*Figure 2. Location map of four transects in Chesapeake Bay.*

portance to oxygen depletion were tested. For example, Tuttle et al. (1987) and Divan and Tuttle (1988) tested the hypothesis that sulfate reduction in the water column was limited by the availability of specific organic substrates. Bell et al. (1988) and Jonas et al. (1988) tested the hypothesis that certain classes of organic compounds, such as amino acids and short chain organic acids, might accumulate near the pycnocline under highly stratified conditions and that this accumulation might be associated with increased bacterial abundance and activity.

The following section summarizes the findings about bacterial behavior, namely, their abundance, production, metabolism and sulfur cycling.

## Bacterial Behavior

### *Bacterial Abundance*

A significant finding of these investigations is the routine occurrence of exceptionally high, sustained bacterial abundances in Chesapeake Bay during spring and summer. In comparison with other estuaries, abundances of such magnitude are large (Ducklow 1982; Jonas 1981; Palumbo and Ferguson 1978; Wright and Coffin 1984). Figure 3, for example, shows that during June, July and August cell densities are usually greater than 7 million cells $ml^{-1}$ and often approach 20 million $ml^{-1}$.

High levels of bacterial abundance occur throughout the region from the Bay Bridge at Annapolis in the north to, at least, the Great Wicomico River in the south. Compared with the midsection of the Bay near the Patuxent River, mean summer abundances were somewhat higher both in the northern portion of the study area (Tuttle et al. 1987b) and the southern portion (Jonas and Tuttle 1990). The wide distribution of elevated bacterial abundances strongly suggests an equally wide distribution of high concentrations of labile organic matter and high primary production rates throughout this region of the Bay.

Bacterial abundance changes seasonally (Figure 3): in the upper mixed layer of the water column in winter, cell densities are about 1 to 2 million cells $ml^{-1}$; in spring, 6-8 million $ml^{-1}$; and in summer 10 to 20 million $ml^{-1}$. Beginning in late August or September, abundances decline and reach a minimum in January or February. The causal significance of the strong association between temperature and bacterial abundance ($r^2$ = 0.72-0.81) (Ducklow et al. 1988) is complicated by complex environmental changes, for example, increased insolation, nutrient input, phytoplankton biomass and production, which also accompany seasonal temperature variations (Tuttle et al. 1987b).

During late spring and summer mean bacterial abundances are often higher along the flanks of the Bay than over the deep main stem channel (Figure 4). In the northern sections of the study area, abundances tend to be highest on the eastern flank (Tuttle et al. 1987a), though in the southern area they tend to be highest on the western margin (Jonas 1987; Jonas and Tuttle 1990). This distribution pattern is probably related to higher phytoplankton production and biomass and higher organic carbon concentrations over the flanks (Malone et al. 1986; Jonas 1987; Malone, this volume).

With more highly stratified waters in summer, bacterial abundance declines with depth (Figure 3), especially when the deep waters are anoxic (Tuttle et al. 1987b). This decline is coupled directly to the location of the pycnocline, with higher abundances above and lower abundances below. In contrast, during late spring, bottom water bacterial abundances can exceed those in the surface waters (Figure 5) (Jonas and Tuttle 1990). These bacterial distributions may be a consequence of earlier accumulations of phytoplankton in the deep water.

The hypothesis that there could be mid-water maxima in bacterial abundance in the vicinity of the pycnocline is supported in observations by Tuttle et al. (1987b) and Jonas and Tuttle (1990). However, zonation of bacteria appears to be quite narrow and it is, therefore, important to sample with high resolution in order to detect the peaks. While the obvious solution to this problem is to collect samples at more depths near the pycnocline, in practical terms this has often been incompatible with other research goals.

In 1987, Jonas was able to incorporate a more intensive depth-

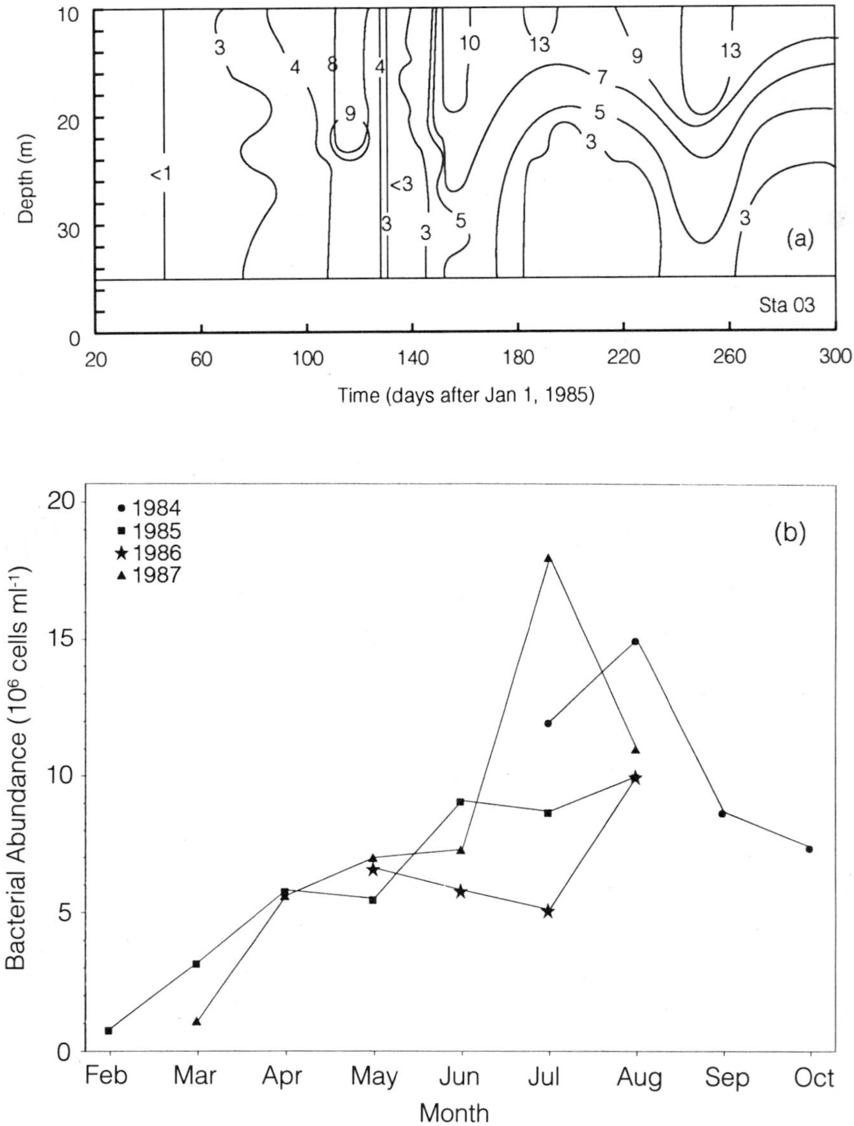


*Figure 3. (a) Time dependent depth profile of bacterial abundance ($10^6$ cells $ml^{-1}$) over the main channel of Chesapeake Bay off the Choptank River (from Tuttle et al. 1987a); (b) mean bacterial abundances in mesohaline Chesapeake Bay, 1984-1987 (from Ducklow et al. 1988).*

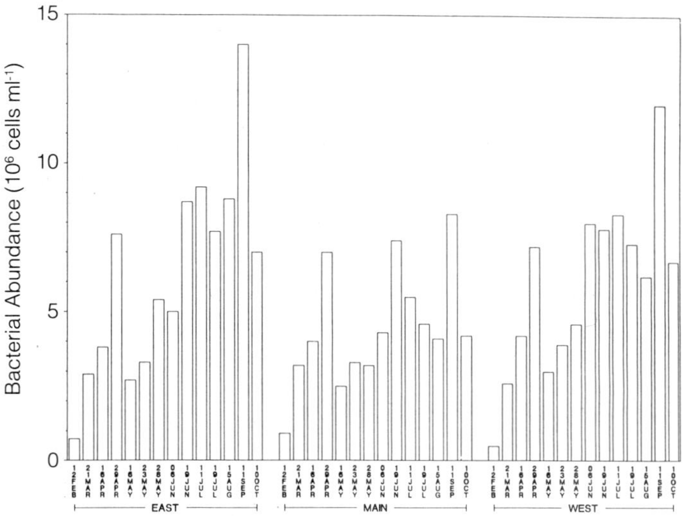


*Figure 4. Mean bacterial abundances over the east and west flanks of the Bay and over the main channel of Chesapeake Bay during 1985 (Tuttle et al. 1987a).*

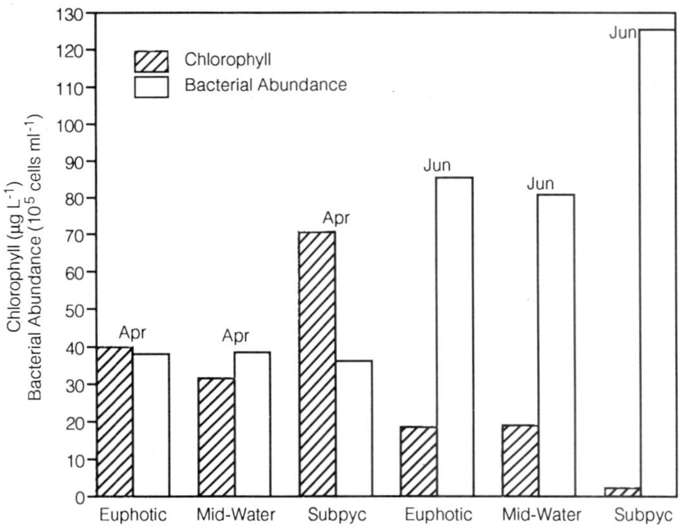


*Figure 5. Depth integrated mean chlorophyll* a *concentrations and bacterial abundance in the euphotic zone (euphotic), between the euphotic zone and the pycnocline (mid-water) and below the pycnocline (subpycnocline) on April 10 and June 9, 1987 at a main channel station off the Great Wicomico River, Virginia.*

profiling sample design into studies between the Patuxent River and the Great Wicomico River. Those investigations showed mid-water maxima of bacterial abundance, as well as bacterial metabolism and concentrations of specific organic compounds (Bell et al. 1988; Jonas et al. 1988a), an indication of the accumulation of materials near this density discontinuity. These findings are most important from a process perspective, because localization of oxygen-demanding organic matter and bacterial metabolic activity (a bacterial plate) likely acts as a biological barrier to the movement of oxygen into the subpycnocline water.

Most of the bacteria in the Bay (often more than 90%) are small, <1 micron (μ) and free living. Because of this small size, large numbers do not necessarily imply great biomass. However, biomass calculations employing reasonable conversion factors (20 μgC/bacterial cell, 70 μgC/μgChl) indicate that bacterial biomass is often 30%, and can exceed 60%, of phytoplankton biomass during late summer when bacterial abundances are maximal (Malone et al. 1986; Tuttle et al. 1987a; Ducklow et al. 1988; Jonas and Tuttle 1990). On a biomass basis, the bacteria are very important members of the heterotrophic community.

If particulate matter, as represented by the phytoplankton, is the principle source of organic matter for these bacteria then bacterial biomass seems sometimes excessive and not sustainable for long. However, dissolved organic matter from the phytoplankton, rather than the particulate phytoplankton itself, may be the form of organic matter directly supporting the bacterial biomass. Continuing investigations in this program have focused, in part, on the specific form and quantity of organic matter which actually fuels microbial metabolism and oxygen consumption. Studies by Ducklow et al. (1989) and Blum and Mills (1989) have been carrying this work on.

### *Bacterial Production*

Bacterial production rates, like bacterial abundances, rise significantly from winter through spring and early summer (Figure 6), roughly from about 50 to 1250 mgC $m^{-2}d^{-1}$ (Malone et al. 1986; Ducklow et al. 1988; Jonas et al. 1988c). These carbon-based values are equivalent to rates of about 20 to 700 million bacterial cells produced in each liter of Bay water each hour. Bacterial production in the euphotic zone is often equivalent to 30% or more of phytoplankton production and occasionally reaches 100% or more of phytoplankton production (Ducklow et al. 1988). As a fraction of endogenous primary production, Chesapeake Bay bacterial production is very high compared to more usual values of 20% or less in other pelagic systems (Hobbie and Cole 1984; Ducklow and Hill 1985). These high production rates indicate the highly active state of the bacterial community and suggest the availability of a large pool of organic matter to support this secondary production.

In spring euphotic zone bacterial production rises rapidly, usually reaching maximal values during June, declines during July and August, but can peak again during late summer (Figure 6b). During winter, bacterial production is uniform throughout the water column, but during late spring and summer euphotic zone production is usually greater than subpycnocline production (Figure 6a) (Tuttle et al. 1987a, b; Jonas and Tuttle 1990). Exceptions to this pattern do occur, however. For example, bottom water production can be as much as 2 to 3 times that in the euphotic zone during spring, and mid-water production peaks near the pycnocline are not uncommon.

Spatially, high rates of production, like abundance, occur throughout the mesohaline Chesapeake Bay. Although the highest rates occur to the north in the region between the Annapolis Bay bridge and the Patuxent River, production is often higher south of the Potomac River than in the region of the Patuxent River. This suggests two foci of biological activity in the Bay (Jonas and Tuttle 1990). The northern focus may be associated with nutrient and/or phytoplankton transport from the upper-Bay while the southern focus may be related to deep-water transport of phytoplankton-derived organic matter from the lower-Bay and influx of nutrients from the Potomac River.

Generally, during summer the highest production values are found over the shallow flanks of the Bay rather than the main channel itself (Figure 7) (Tuttle et al. 1987a). Especially in spring

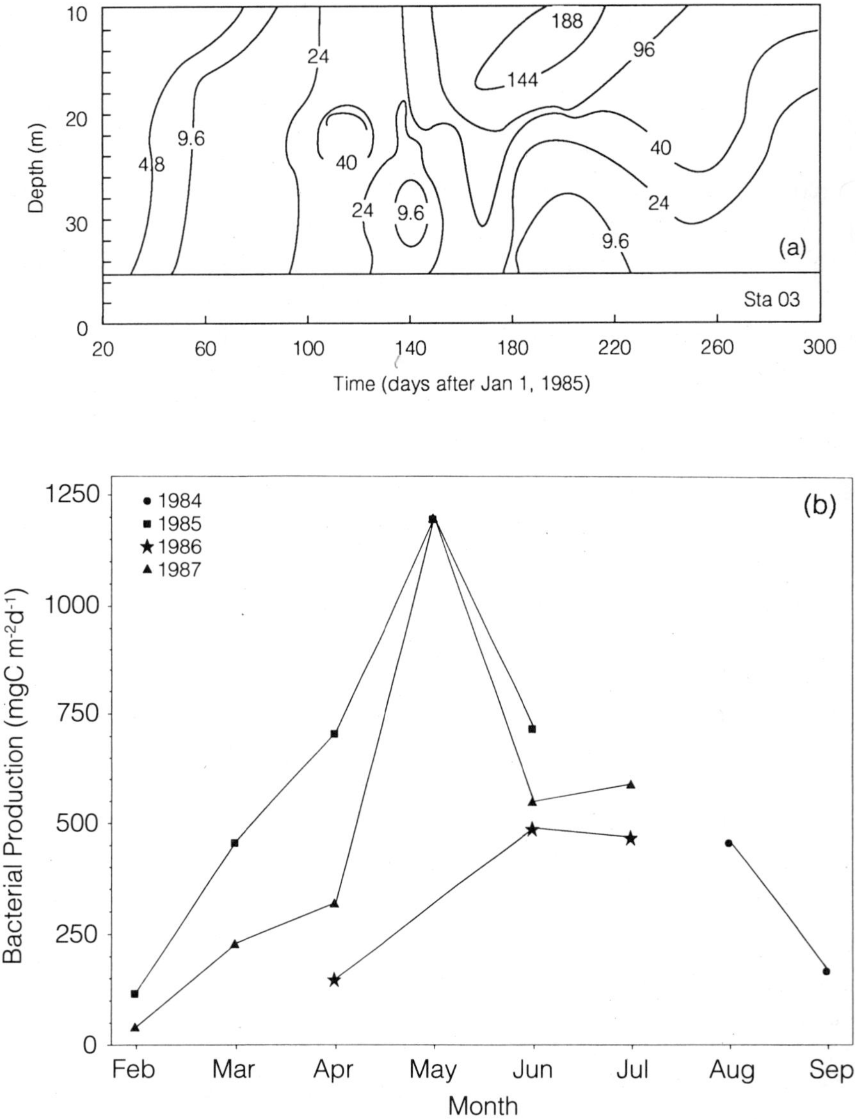


*Figure 6. (a) Time dependent depth profile of bacterial production over the main channel of Chesapeake Bay off the Choptank River (Tuttle et al. 1987a); (b) mean bacterial production in mesohaline Chesapeake Bay, 1984-1987 (Ducklow et al. 1988).*

and summer, highest production values are found along the western flank. This has been particularly apparent along the Great Wicomico River transect in the mid-Bay south of the Potomac River (Jonas 1987). The varying distribution pattern of bacterial production implies a rather complex and finely zoned distribution of the organic matter which supports that production. Evidence suggests that the bacterial production and metabolism in this system are most directly dependent on dissolved organic matter in the water column (Bell et al. 1988; Jonas et al 1988; Jonas and Tuttle 1990). An analysis by Ducklow et al. (1988) and Peele (1988) indicates that phytoplankton, under a set of specifically defined steady state conditions, could support bacterial production at a maximal rate of about 40% of phytoplankton production. They argue that greater relative

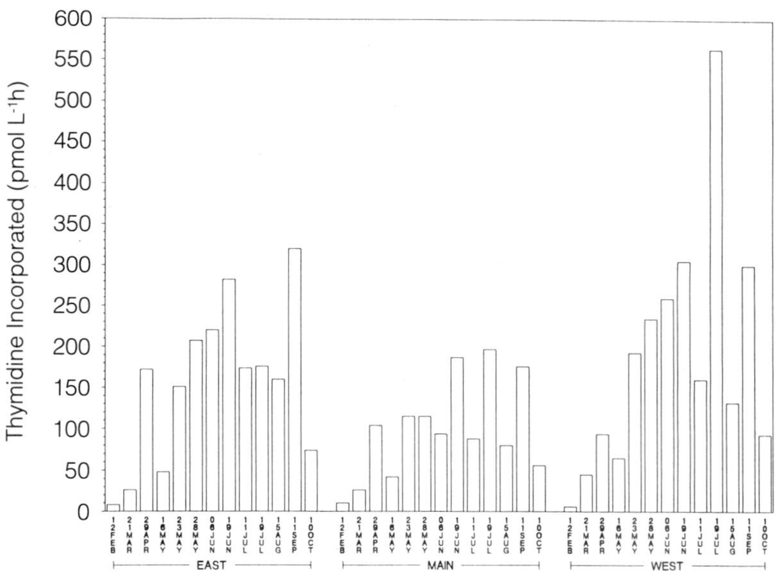


*Figure 7. Mean bacterial production over the east and west flanks of the Bay and over the main channel of Chesapeake Bay during 1985 (Tuttle et al. 1987a).*

production rates would necessitate either external organic carbon sources or a diminished bacterial growth rate. If phytoplankton production measurements do not account for dissolved organic production, then the distribution of bacterially labile organic carbon concentrations should be measured directly (see below) in order to more accurately compare primary and secondary production.[1]

### *Bacterial Metabolism*

Bacterial metabolism of several classes of dissolved organic compounds — including glucose, amino acids and organic acids — is also high compared with other ecosystems (Bell et al. 1988; Jonas et al. 1988a). Data indicate that monosaccharides, probably glucose, are likely the preferred carbon source supporting bacterial metabolism in Chesapeake Bay. Glucose turnover rates in late spring, compared with winter lows of a few percent per hour, can reach 50% $h^{-1}$ and are sustained throughout the summer at 20% $h^{-1}$ (Figure 8a). Amino acid pool turnover rates in spring are slower than glucose rates (Figure 8b) but can reach 20% $h^{-1}$ and generally follow glucose turnover rather closely throughout the year.

Data from three annual cycles indicate that bacterial metabolism, which is elevated throughout the mesohaline Bay, is uncoupled from bacterial production temporally. Metabolic activity peaks in April or May while production peaks in June. Geographically, however, metabolism, like production, is greater over the flanks of the Bay than over the main channel (Figure 8c) and is usually slightly greater toward the northern and southern boundaries of this region.

---

[1] There is some concern about the appropriate use of the thymidine incorporation method for estimating bacterial production (Karl 1982; Lovell and Konopka 1985; Robarts et al. 1986). One of the most significant concerns is the possibility that macromolecules other than DNA become radiolabelled during the incubation procedure. As it is applied here, the technique is likely to yield acceptable estimates of bacterial production in Chesapeake Bay, especially when the water is not anoxic. Under aerobic conditions, most of the titrated thymidine is apparently incorporated into DNA. Given the need for accurate estimates of bacterial secondary production, further investigation of the specific applicability of the techniques to Bay samples and determinations of appropriate cell and carbon conversion factors are certainly warranted.

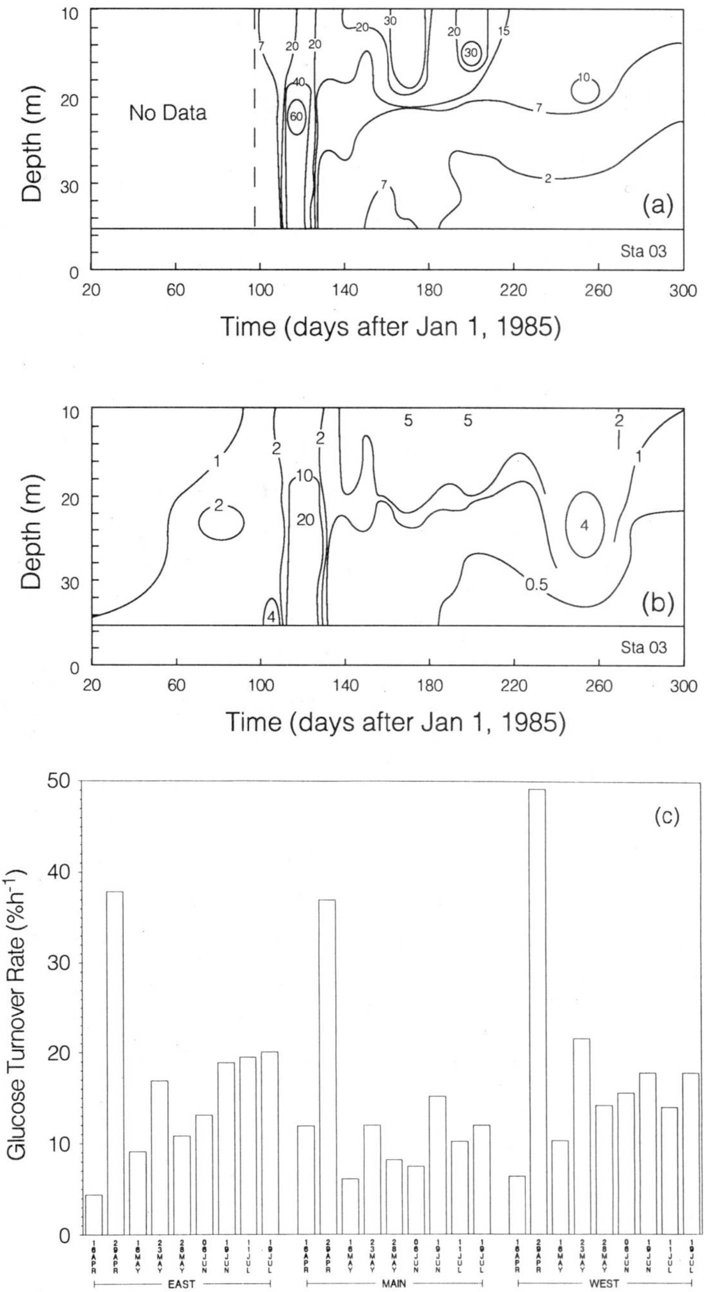


*Figure 8. Time dependent depth profiles of (a) glucose and (b) amino acid turnover rates (% $h^{-1}$)over the main channel of Chesapeake Bay off the Choptank River; (c) mean glucose turnover rates over the east and west flanks and over the main channel of Chesapeake Bay during 1985 (from Tuttle et al. 1987a).*

To calculate actual rates of metabolism of specific substrates both turnover rate and concentrations must be known. In 1987, free dissolved glucose concentrations were estimated to be in the 1 to 2 μmolar range in the southern mesohaline Chesapeake Bay and free amino acid concentrations ranged from about 0.2 to 1.2 μmolar (Bell et al. 1988; Jonas et al. 1988a). Average metabolic rates were about 225 nmol $L^{-1}h^{-1}$ for glucose and about 20 nmol $L^{-1}h^{-1}$ for amino acids. The data indicate that organic pool turnover rates actually increase as substrate concentrations increase, which implies that the bacterial community is poised to respond quickly to take advantage of increased nutrient concentrations by expanded metabolic activity. This response could be accomplished either by increasing cell abundance or increasing cell specific rates of metabolic activity. Both responses probably take place in the Bay bacterial community.

Bacterial metabolic activity peaks in the vicinity of the pycnocline under some conditions (Bell et al. 1988; Jonas et al. 1988a). During summer, when the water column is stratified, amino acid metabolism can be 3 to 7 times greater at the pycnocline than it is above or below (Figure 9). This increase in activity is associated with a 5 to 10 fold increase in amino acid concentration at the pycnocline, but no marked increase in glucose concentration. Conservatively, bacterial amino acid and glucose metabolism alone in the water column near the pycnocline could result in a 0.7 mg $L^{-1}d^{-1}$ demand on dissolved oxygen. Thus, the bacteria could effectively "cap" the subpycnoclinal region and prevent diffusive reaeration.

### *Bacterial Sulfur Cycling*

In addition to aerobic heterotrophic processes which consume oxygen (Equation 1), the biogeochemical cycling of sulfur also leads to a major oxygen demand in the water column. Under anoxic conditions, sulfate reducing bacteria carry out anaerobic respiration using sulfate as a terminal electron acceptor (Equation 2) and produce hydrogen sulfide as a byproduct:

$$\text{Aerobic Respiration: } [CH_2O] + O_2 \rightarrow CO_2 + H_2O \qquad (1)$$

Sulfate Reduction: $2[CH_2O] + SO_4^= \rightarrow S^= + 2CO_2 + 2H_2O$ (2)
(Anaerobic Respiration)

Sulfide Oxidation: $S^= + 2O_2 \rightarrow SO_4^=$ (3)

In turn the hydrogen sulfide can be oxidized back to sulfate and/or other sulfur intermediates (Equation 3), in the presence of dissolved oxygen, by both microbial and chemical processes. In the Bay's mesohaline region, hydrogen sulfide production likely occurs continuously in the sediments, but sulfide only accumulates in the water column when it is anoxic. Under those conditions, sulfide can often be detected just beneath the pycnocline and increases in concentration with depth to the bottom (Figure 10) (Tuttle et al. 1987a). In the Bay's mesohaline region, water column sulfide concentrations reach a maximum of about 8-15 µmolar. Stoichiometrically, this sulfide concentration would demand 0.25-0.50 $mgO_2\ L^{-1}$.

Sulfide is produced primarily in the bottom sediments (Tuttle et al. 1987a). Currently available data suggest that only about 1% of the sulfide is produced in the water column when it is anoxic, and that water column sulfate reduction is likely limited by low concentrations of specific organic substrates required by sulfate reducing bacteria (Divan and Tuttle 1988; Tuttle et al. 1987). Though the rate of sediment sulfate reduction is strongly correlated with temperature in spring and summer, peak rates occur in early fall, probably in response to organic inputs from late summer and fall phytoplankton production (Tuttle et al. 1987a).

When the water column is oxygenated, sediment sulfate reduction continues but the sulfide is oxidized at the sediment water interface. Therefore, its influence on water column oxygen concentrations is physically limited. However, once the water column becomes anoxic, sulfide diffuses upward at a rate controlled by both the rate of production of sulfide and rate of reaeration and can constitute a major oxygen demand at the pycnocline. Under summer conditions measured sulfide oxidation in the vicinity of the pycnocline results in mean oxygen consumption rates of 9 $mgO_2\ L^{-1}\ d^{-1}$ (Tuttle et al. 1987b). Although this demand is limited to the region immediately beneath the pycnocline, it is comparable to the

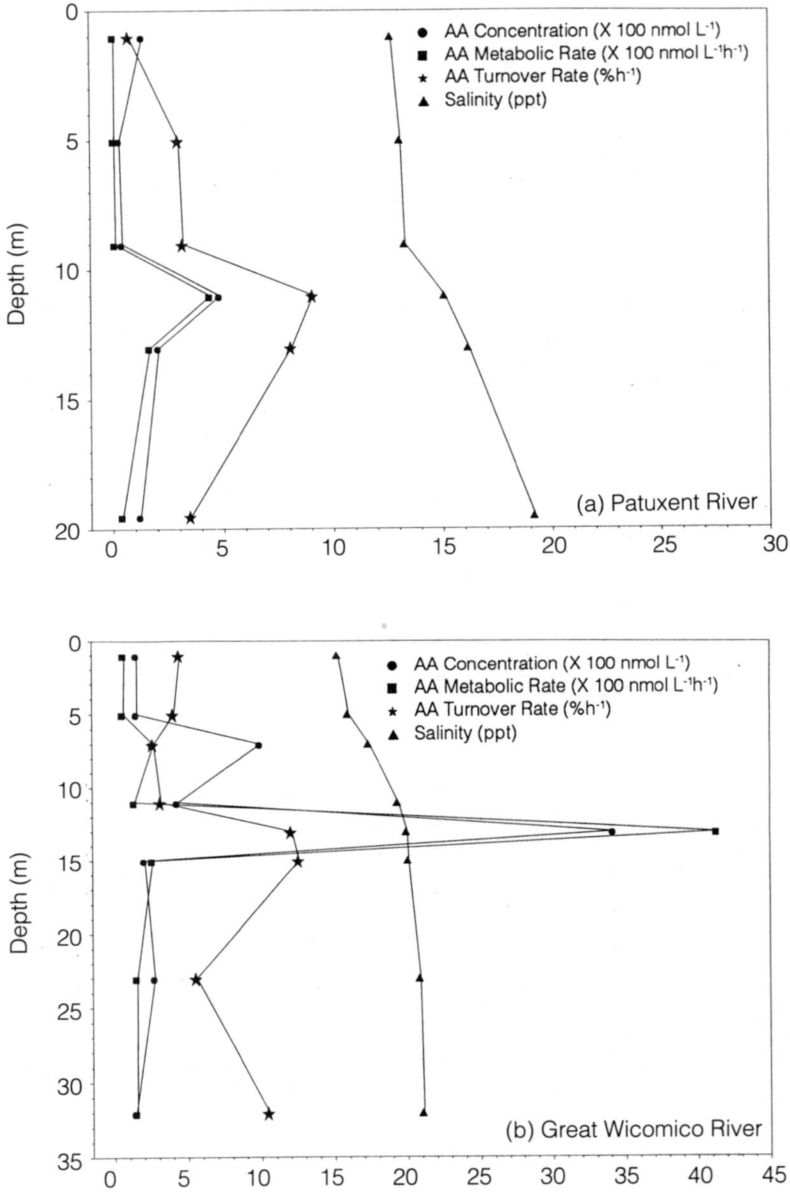


*Figure 9. Vertical distributions of amino acid concentrations, amino acid turnover rates and amino acid metabolic rates at main stem stations off (a) the Patuxent River and (b) the Great Wicomico River on July 30, 1987.*

measured (under aerobic conditions) aerobic heterotrophic demand in the water column beneath the pycnocline.

## *Modeling the Bacterial Role in Oxygen Consumption*

Models that detail the processes influencing oxygen consumption in the Bay should emphasize the importance of(1) aerobic heterotrophic bacterial metabolism, (2) sulfide oxidation and (3) accumulation of bacterially labile organic matter in the water column.

Accumulations in the vicinity of the pycnocline may be especially important. Evidence to date indicates that during spring and summer, microbial aerobic heterotrophic metabolism throughout the water column results in the progressive depletion of oxygen beneath the pycnocline during spring and summer. Once anoxic conditions develop, sulfide oxidation at the pycnocline consumes oxygen both biologically and chemically and is a key component in

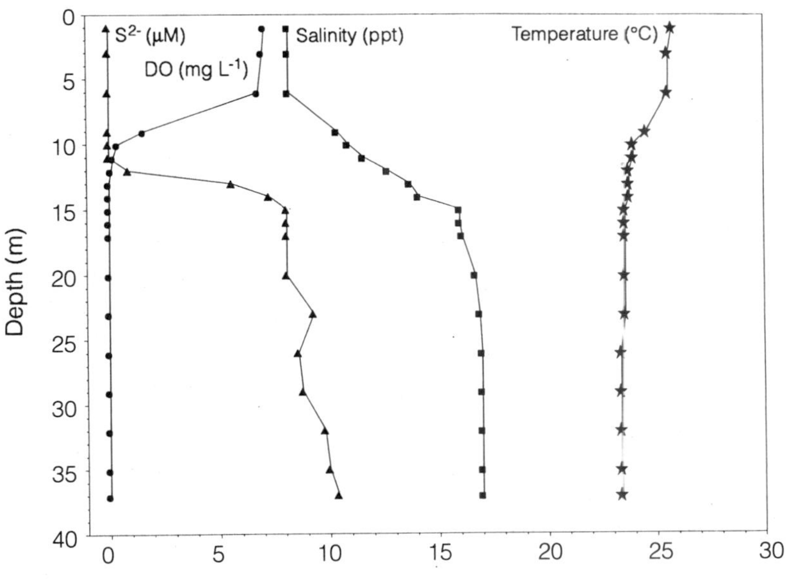


*Figure 10. Vertical distribution of dissolved oxygen, sulfide, temperature and salinity at a main stem station off the Patuxent River on July 26, 1984 (from Tuttle et al. 1987a).*

maintaining anoxic conditions. In addition, organic matter (i.e., amino acids) and fermentation products (i.e., acetate) accumulate at the pycnocline and fuel very high local rates of bacterial metabolism and oxygen consumption (Tuttle et al. 1987a).

Several lines of evidence support a conclusion that the bacteria are responsible for the oxygen consumption.

1. Measured net oxygen consumption rates in the water column are about 1 to 1.5 $mgO_2 L^{-1}d^{-1}$ (Tuttle et al. 1987a,b).

2. Calculated oxygen consumption rates based on the measured bacterial abundances and biomass specific oxygen consumption rates for bacteria are also about 1 to 1.5 $mgO_2 L^{-1}d^{-1}$.

3. Size fractionation experiments indicated that 50 to 95% of the oxygen consumption in surface waters, and 70 to 100% in bottom water, could be accounted for by organisms smaller than 3μ (Kemp et al. 1987). Although this could include eukaryotic microorganisms such as protozoans, bacteria likely dominate this size range.

Sampou et al. (1988) reported that during May, when oxygen is rapidly depleted, 60 to 70% of total biological oxygen consumption is attributable to planktonic respiration; data from 1986 and 1987 indicate that bottom water oxygen consumption is due principally to organisms smaller than 3μ. There was a strong linear relationship ($r^2 = 0.76$, slope = 1.1) between total water column respiration and respiration in 3μ filtered water. If the bacteria are actually responsible for the oxygen consumption, a strong linear relationship should exist between bacterial abundance or activity and the rate of oxygen depletion. Regression of water column oxygen consumption on bacterial abundance for 1984 and 1985 ($y = 0.0045X + 0.0066$; $r^2 = 0.90$) clearly supports the importance of bacterially mediated oxygen consumption (Figure 11) (Tuttle et al. 1987a). Since there is a relatively strong correlation of oxygen consumption on temperature (upper water column $r^2$ = 0.69, $Q_{10}$ = 2.7[2]; lower water column $r^2$ = 0.23, $Q_{10}$ = 1.6) (Sampou et al. 1988), the relationship shown in Figure 11 implies some kind of adaptation within the microbial community such that bacterial cell specific oxygen consumption is relatively unchanged throughout the year.

### *Organic Matter in the Water Column —What to Measure*

Any attempt to model the physical, chemical and biological interactions leading to oxygen depletion must naturally account for all of the most important factors influencing this process. One such factor, the pool of organic matter which actually supports the microbial metabolism, has not previously been measured directly. Estimates of particulate organic matter, such as chlorophyll *a*, particulate organic carbon and particulate organic nitrogen, which are normally used as indicators of water quality, do not account for the dissolved organic matter which is likely the primary form upon which the microbes depend (Bell et al. 1988; Ducklow et al. 1988; Jonas et al. 1988a,b; Jonas and Tuttle 1990). Analytical measures of total dissolved organic carbon are inappropriate because they are dominated by refractory "yellow" material (Gelbstoffe) which is not microbially labile.

Therefore, in conjunction with this research, measurements were taken of both total and dissolved biochemical oxygen demand (BOD) and dissolved, free amino acid and monosaccharide concentrations (Bell et al. 1988; Jonas et al. 1988; Jonas and Tuttle 1990). The 5-day BOD measurement provided an empirical estimate of the amount of bacterially labile organic matter independent of its chemical composition, whereas analytical determinations of amino acids and carbohydrates were used to identify and quantitate the specific classes of organic matter supporting bacterial metabolism.

Biochemical oxygen demand in the Bay ranges from less than 0.5 mg $L^{-1}$ to more than 7 mg $L^{-1}$ on an annual basis and varies

---

[2] The $Q_{10}$ value is a measure of the metabolic rate increase for each 10°C in temperature. A $Q_{10}$ of 2 means that the metabolic rate doubles with increase of 10°C.

significantly from season to season. During summer, maximum values occur in association with phytoplankton blooms in surface waters along the western side of the Bay, especially near the Annapolis Bay Bridge and in the area south of the Potomac River (Jonas 1987). In early spring, however, BOD is highest in bottom water, averaging 3-4 mg $L^{-1}$. The inverse relationship with depth mirrors spring chlorophyll distributions. Early spring surface BOD is higher in the southern mesohaline Bay than to the north, but during late spring the reverse is true. In spring — but not summer — mid-water peaks in BOD are frequent (Jonas and Tuttle 1990).

During spring, most BOD is particulate in form throughout the water column (Jonas et al. 1988b, Jonas and Tuttle 1990). In summer, however, dissolved material dominates the BOD especially below the pycnocline, accounting for 50-60% of the total BOD throughout the water column and 80% in the bottom water. Obviously, if only particulate organic matter is measured, a large fraction of the material supporting the oxygen demand will be missed. This uncertainty could seriously limit the accuracy of environmental models designed to predict Chesapeake Bay responses to changes in nutrient inputs.

Free dissolved carbohydrates and amino acids are present in high concentrations in Bay water, and during summer may account for about 50% of the total BOD (Table 1) (Jonas et al. 1988a). These substrates may be even more important than their static concentrations suggest because high bacterial turnover rates (e.g., 50% $h^{-1}$ for glucose) imply that the pools are rapidly renewed so as to maintain these high concentrations. The dissolved free carbohydrates appear to be a major substrate fueling bacterial oxygen consumption throughout the water column. However, under highly stratified conditions and subpycnoclinal anoxia, amino acids, and probably short chain organic acids, can accumulate near the pycnocline and become proportionately much more important (Jonas et al. 1988a,b).

Although phytoplankton are probably the source of the organic matter supporting the bacterial community and oxygen consumption (Malone et al. 1986), a strong linear relationship is not

Table 1. Oxygen demand due to measured biochemical oxygen demand (BOD), soluble biochemical oxygen demand (FBOD), free dissolved monosaccharides (glucose) and free dissolved amino acids (AA), and oxygen consumption due to monosaccharides and amino acids under summer conditions in 1987 at two mainstem Chesapeake Bay stations.

| | BOD ($mg\ L^{-1}$) | FBOD ($mg\ L^{-1}$) | Glucose ($mg\ L^{-1}$)[1] | Amino Acid ($mg\ L^{-1}$)[1] | Oxygen Consumption ($mgO_2\ L^{-1}d^{-1}$) | |
|---|---|---|---|---|---|---|
| | | | | | GLU | AA |
| Patuxent Station a[2] June - Aug. | 1.36 | 0.77 | 0.16 | 0.05 | nr | nr |
| Great Wicomico Station b[3] June - Aug. | 1.08 | 0.76 | 0.16 | 0.11 | nr | nr |
| Combined Station Mean June - Aug. | 1.20 | 0.76 | 0.16 | 0.08 | 0.30 | 0.07 |
| Combined Station Mean August at Pycnocline | 1.16 | 1.01 | 0.25 | 0.20 | 0.50 | 0.20 |
| Combined Station Mean August Excluding Pycnocline | 1.57 | 0.94 | nr | 0.05 | nr | nr |

[1] Measured concentrations were converted to oxygen demand equivalents assuming that all carbon is oxidized to $CO_2$. Oxygen consumption rates were calculated using measure glucose and amino acid metabolic rates. Oxygen consumption was calculated based on amino acid metabolic rates. Oxygen consumption was calculated based on measured metabolic rates for the specific substrates.

[2] Station a located over main channel opposite Patuxent River.

[3] Station b located over main channel opposite Great Wicomico River.

evident between measures of phytoplankton biomass (chlorophyll *a*) and BOD ($r^2 = 0.24$). A somewhat stronger relationship is evident among particulate organic carbon or particulate organic nitrogen and BOD ($r^2 = 0.54$ and 0.64, respectively); although this accounts for 60% of the variation, it is clear that measures of particulate organic matter do not explain all of the variation in organic carbon fueling bacterial metabolism. We suggest that measurement of BOD and dissolved BOD should be included in the suite of variables monitored in Bay waters and that these data be considered in water quality monitoring and modeling efforts.

Studies by Lagara and Zieman (1988) and Zieman and Macko (1988) support the hypothesis that phytoplankton are the principal source of organic matter in the Bay. They used a technique which tracks the biological source of organic carbon, nitrogen and sulfur by quantifying differential enrichment or depletion of naturally occurring stable isotopes of each element. Their data indicate that the stable isotope enrichment pattern in Chesapeake Bay biota is consistent with a phytoplankton origin for the organic matter but not consistent with an upland terrestrial or submerged aquatic macrophyte source.

## *Ecological Role of Bacterioplankton*

Fine scale zonation of organic matter, bacteria and their associated processes in the water column appears to be characteristic of the mesohaline Chesapeake Bay during the summer. This pattern corresponds to the "bacterial plate" concept in which functionally specialized bacterial communities are arranged in layers in the vicinity of a density discontinuity, especially when anoxic conditions develop. Given the great diversity in metabolic potential exhibited by bacteria, expectations are that the various habitats occurring throughout the water column would be exploited by opportunistic bacteria. The linkage between these bacterial communities and water quality conditions in Chesapeake Bay should be of special concern to water quality management for two important ecological reasons. First, the accumulation of organic matter at the pycnocline results in a biological barrier to reaeration of the deep water. Second, there is a critical biological distinction between hypoxic conditions in the water column and true anoxia.

### *Accumulation of Organic Matter at the Pycnocline*

Recent data (Bell et al. 1988; Jonas et al. 1988b) would suggest that bacterially labile organics can derive from below, as well as above, the pycnocline. Specifically, short chain organic acids, which are probably fermentation end products, fuel high rates of bacterial metabolism in the immediate vicinity of the pycnocline. These products probably accumulate only under anoxic conditions and are metabolized at the narrow interface between oxic and anoxic zones. From a management perspective a critical result of this ecological interaction is that bacterial oxygen consumption actually limits the rate of oxygen diffusion to the deep waters by reducing the oxygen concentration gradient near the pycnocline. It seems reasonable to suggest that this kind of fine scale process could materially affect oxygen dynamics in the Bay.

### *Significant Differences between Anoxia and Hypoxia*

While hypoxia itself can and does have dire consequences, anoxia — once it occurs in the system — alters conditions to such an extent that, in a major way, bacterial processes take control of the ecoysystem. Under anoxic conditions end products of fermentation and anaerobic respiration can accumulate and fuel increased rates of oxygen consumption. This positive feedback mechanism helps to maintain the dissolved oxygen deficit. Similarly, hydrogen sulfide and possibly other reduced sulfur compounds can accumulate in the water column and constitute an oxygen demand. It is important to remember that as long as even a small amount of oxygen remains, the ecological transition will not occur.

### *Bacterial Processes and Trophic Dynamic Change*

Just what consequences high levels of bacterial activity have for the Bay's ecosystem are still open to question. One hypothesis is that there has been a shift in the Bay's trophic structure from a metazoan chain, in which primary production is grazed by primary consumers such as zooplankton, to a microbial chain, in which primary production is consumed by bacteria (Tuttle et al. 1987a). Though speculative, there is evidence to support such an hypothesis (Malone, this volume).

A comparison of data from 1984 — a year of above average rainfall and runoff — and 1985 — a year of below average rainfall — is instructive in this regard. While widespread anoxia occurred during the summer of 1984, anoxia was very limited both in duration and extent in 1985. Despite these differences, mean bacterial abundance, both above and below the pycnocline, was very similar in both years (Table 2) (Tuttle et al. 1987a). In contrast, bacterial production was about 40% greater in 1985 than in 1984. Bacterial turnover rates calculated from these measurements were about twice as great in 1985 than in 1984. These data suggest that in 1985, when oxygen was present, the bacteria must have been grazed or otherwise removed twice as fast as in 1984. Perhaps anoxic conditions and/or hydrogen sulfide toxicity in 1984 may have inhibited the bacterioplankton grazers. In essence, such a scenario reflects a positive feedback loop: as anoxia persists longer, fewer bacterivores are active and bacteria are free to consume more oxygen. The bacterial community in this case might be considered analogous to a secondary biological wastewater treatment system where metabolically active bacteria are used to reduce organic loading as rapidly as possible. The analogy, however, does break down: unlike the natural system, water is aerated in a waste treatment plant to prevent anoxia.

It appears that a large portion of the phytoplankton production is processed directly by the bacteria rather than by higher organisms. The loss of a substantial portion of benthic filter feeding populations, particularly oysters, which previously removed phytoplankton from the water column, provides one possible scenario supporting this hypothesis. As the natural abundance of filter feeding oysters is reduced through harvesting, poor water quality or disease, phytoplankton populations could increase; this releases more microbially labile organic matter to the water column, which supports increased bacterial abundances and contributes to even more oxygen depletion.

Table 2. Phytoplankton and bacterial parameters under summer conditions in 1984 and 1985 in the mesohaline region of Chesapeake Bay.

| | | Euphotic Zone | Aphotic Zone |
|---|---|---|---|
| Phytoplankton Production | (1984) | 1.20 | — |
| ($gC\ m^{-2}d^{-1}$) | (1985) | 1.37 | — |
| Phytoplankton Biomass | (1984) | 56.1 | — |
| ($mgChl\ m^{-2}$) | (1985) | 36.6 | — |
| Bacterial Production[1] | (1984) | $2.9x10^{8}$ | $1.9x10^{8}$ |
| ($cells\ L^{-1}h^{-1}$) | (1985) | $5.3x10^{8}$ | $3.1x10^{8}$ |
| Bacterial Abundance | (1984) | $1.4x10^{10}$ | $0.8x10^{10}$ |
| ($cells\ L^{-1}$) | (1985) | $1.1x10^{10}$ | $0.7x10^{10}$ |
| Bacterial Turnover | (1984) | 0.5 | 0.5 |
| ($d^{-1}$) | (1985) | 1.2 | 1.1 |
| $O_2$ Consumption | (1984) | 1.5 | 1.0 |
| ($mg\ L^{-1}d^{-1}$) | (1985) | 1.3 | 1.0 |
| $O_2$ Consumption[2] | (1984) | 1.6 | 1.1 |
| ($mg\ L^{-1}d^{-1}$) | (1985) | 1.3 | 0.9 |

[1] Calculated as $2x10^{18}$ cells produced per mol thymidine incorporated.

[2] Calculated from bacterial abundances and relationship (shown in Figure 11).

Newell (1988) and Jonas et al. (1989) have hypothesized that by increasing the abundance of filter feeders such as oysters, phytoplankton biomass in the water column could be reduced. The filter feeders would metabolize some of the phytoplankton production directly and deposit substantial amounts on the bottom as feces or pseudofeces, where it likely would exert its oxygen demand only slowly. Based on estimates of standing stocks in the Bay, oysters at the end of the 19th century could have filtered all of the water in Chesapeake Bay once every four days (Newell 1988). Current oyster biomass would require many months to accomplish the same filtration. It seems possible that anthropogenic and disease related reductions of the benthic biota may have contributed substantially to the hypothesized trophic shift toward a phytoplankton/bacterioplankton system.

Currently, there is no clear indication as to the ultimate "resource" productivity of a bacterial, secondary producer based food web. It is possible that bacterivores such as protozoans could provide a linkage between the bacteria and higher organisms and, thus, bacteria could form the base of a productive food web. However, due to increased trophic transfers, secondary producer based systems are less efficient than primary producer/grazer systems. This inefficiency suggests that bacteria do not act as the base of an important food web, but rather may act as a sink for carbon and energy (Ducklow 1983; Ducklow et al. 1986). Even if the bacteria do form

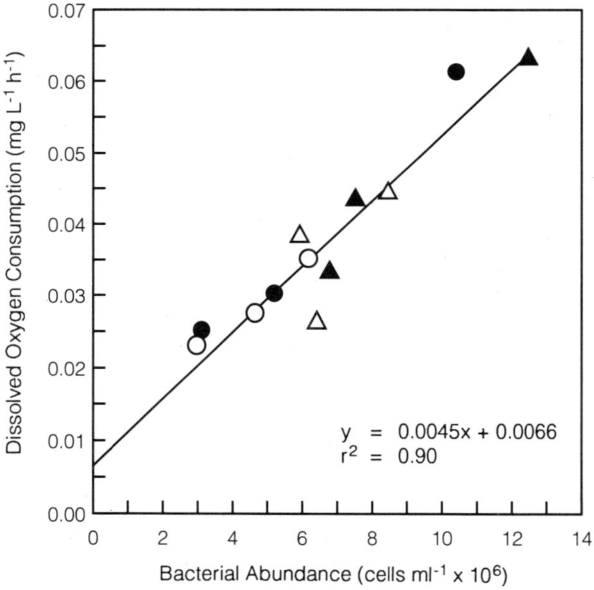


*Figure 11. Relationship between water column oxygen consumption and bacterial abundance. Symbols: (▲) euphotic zone means; ( ● ) below euphotic zone means for periods Aug 20-Sep 11, Sep 14-Oct 3 and Oct 5-Nov 2, 1984; ( △ ) euphotic zone means; ( ○ ) below euphotic zone means for periods Feb 12-Apr 17, Apr 29-June 7 and June 19-Sep 12, 1985 (from Tuttle 1987a).*

the base of a valuable food web, the intermediate costs of hypoxia and anoxia are probably too large to be compensated by this food web. Further investigations of the relationships among bacteria and their consumers are clearly needed to settle these questions.

## Research and Management Needs

While it is unknown just how abundant or metabolically active bacteria were in the past history of Chesapeake Bay, two important outcomes of these investigations have been (1) the realization of the importance of the heterotrophic microbial component in relation to water quality and oxygen dynamics and (2) the recognition that an understanding of the microbial role in the Bay's trophic dynamics is essential for its management. In addition, these studies have provided the opportunity for significantly improved understanding of the more basic interactions among nutrients, phytoplankton and bacteria in estuarine environments. Nevertheless, that understanding is far from complete. More sophisticated management will require studies that further develop knowledge of the following:

1. Continue studies of nutrient, phytoplankton, bacterial interactions in the upper Bay and major tributaries. Do the same relationships exits in these environments? How do tributaries influence the main stem Chesapeake Bay?

2. Focus investigations on the specific sources of bacterially labile organic matter. Test specifically the nature and quantity of organics released by phytoplankton and grazers.

3. Evaluate the concentrations of important bacterially labile organic pools.

4. Examine trophic interactions which may have exacerbated the low dissolved oxygen problem. What effect does low dissolved oxygen have on food web functioning? What

influence do bacterivores have on bacterial processes, including rates of oxygen consumption?

5. Probe the possible existence of an ecologically important bacterially based food web in the Bay. Does such a food web function in an energetically important way to support harvestable resources?

Several of these areas are currently being investigated with support of the National Oceanic and Atmospheric Administration through the joint Maryland and Virginia Sea Grant College programs, Environmental Effects Research on Chesapeake Bay. Specifically, Ducklow et al. (1989) are investigating the role of bacteria in decomposition of phytoplankton-derived detritus, and the relative importance of particle associated bacterial processes in the upper- and mesohaline Bay. In contrast, Blum and Mills (1989) are evaluating the nature and quantity of dissolved organic matter released as fermentation byproducts from the anoxic zone in the Bay's bottom. Both particulate and dissolved carbon contribute to oxygen demand and, therefore, one overall goal of these efforts will be to assess the relative contribution of each to oxygen depletion in the Bay.

Obviously, knowledge of trophic transfer rates of bacterioplankton and phytoplankton is important in understanding the food web structure of the Bay. Previous work suggests that the Bay ecosystem may have shifted toward a bacterioplankton dominated system, in part, due to loss of phytoplankton consumers such as oysters. In a field study, Jonas et al. (1989) are investigating the hypothesis that reintroduction of large abundances of filter feeding oysters, through a rafted oyster aquaculture approach, will significantly reduce bacterioplankton, phytoplankton and organic carbon in the water column. McManus (1989), recognizing the importance of bacterioplankton in the Chesapeake, is conducting studies to determine the rates of bacterivory by protists and micro-metazoans.

Additionally, it continues to be important to support the development and testing of specific methods for probing the "invisible" microbial world. This discussion began with the suggestion that the new understanding of the importance of microbial heterotrophs was based, in part, on technical innovations and improvements in recent years. It seems plausible to suggest that further refinements of our knowledge will come from even more sophisticated approaches. Some of these, for example, image analysis and genetic probes, are an immediate prospect while others may only be in early stages of development. Despite the intense competition for the economic resources to support the Bay efforts, it may well be important to support focused efforts designed to improve analytical approaches.

## References

Atlas, R.M. and R. Bartha. 1987. Microbial Ecology: Fundamentals and Applications (2nd Edit.) The Benjamin/Cummings Publishing Co., Inc., Menlo Park, California.

Bell, J.T., D. Gluckman, R.B. Jonas and J.H. Tuttle. 1988. Fine scale zonation in microbial turnover of labile dissolved organic substrates in Chesapeake Bay. Presented at the American Geophysical Union, American Society for Limnology and Oceanography Ocean Sciences Meeting, New Orleans, Jan. 1988.

Biggs, R.B. and D.A. Flemer. 1972. The flux of particulate carbon in an estuary. Mar. Biol. 2:11-17.

Blum, L.K. and A.L. Mills. 1989. The role of dissolved organic constituents from anaerobic decomposition in oxygen dynamics in the water column, pp. 54-72. In: Research Proposal: Environmental Effects Research on Chesapeake Bay, Dissolved Oxygen and Related Processes. Maryland and Virginia Sea Grant College Programs.

Carpenter, J.H. and D.G. Cargo. 1957. Oxygen requirement and mortality of the blue crab in the Chesapeake Bay. Tech. Rept. No. 13. Reference #57-2. Chesapeake Bay Institute, The Johns Hopkins University, Baltimore.

Carpenter, J.H. 1965. The Chesapeake Bay Institute technique for the Winkler dissolved oxygen method. Limnol. Oceanogr. 10:141-143.

Divan C.L. and J.H. Tuttle. 1988. Water column sulfate reduction in the Chesapeake Bay and Potomac River estuaries. Presented at the American Geophysical Union, American Society for Limnology and Oceanography Ocean Sciences Meeting, New Orleans, January 1988.

Ducklow, H.W. 1982. Chesapeake nutrient and plankton dynamics. I. Bacterial biomass and production during spring tidal destratification in the York River, Virginia, estuary. Limnol. Oceanogr. 27:651-659.

Ducklow, H.W. 1983. The production and fate of bacteria in the ocean. Bioscience 33:494-501.

Ducklow, H.W., D.A. Purdie, P.J.LeB. Williams and J.M. Davies. 1986. Bacterioplankton: A sink for carbon in a coastal plankton community. Science 232:865-867.

Ducklow, H.W. and S.M. Hill. 1985. The growth of heterotrophic bacteria in the surface waters of warm core rings. Limnol. Oceanogr. 30:239-259.

Ducklow, H.W., E.A. Peele, S.M. Hill. and H.L. Quinby. 1988. Fluxes of carbon, nitrogen, and oxygen through estuarine bacterioplankton. In M.P. Lynch and E.C. Krome (eds.) Understanding the Estuary: Advances in Chesapeake Bay Research. Proceedings of a Conference, 29-31 March Baltimore, Maryland. Publication No. 129, CBP/TRS 24/88. Chesapeake Research Consortium, Gloucester Point, Virginia.

Ducklow, H., M. Fletcher and P.C. Griffith. 1989. Trophic dynamics in Chesapeake Bay: Ecological and biochemical approaches to the study of flows of carbon from phytoplankton to bacterioplankton, pp 111-139. Research Proposal: Environmental Effects Research on Chesapeake Bay, Dissolved Oxygen and Related Processes. Maryland and Virginia Sea Grant College Programs.

Environmental Protection Agency (EPA). 1982. Chesapeake Bay Program technical studies: A synthesis. U.S. Environmental Protection Agency, Washington D.C.

Hobbie, J.E. and J.J. Cole. 1984. Response of a detrital food web to eutrophication. Bull. Mar. Sci. 35:357-363.

Jonas, R.B. 1981. Salt marsh microbial community responses to acute and long-term copper stress. Doctoral dissertation, University of North Carolina.

Jonas, R.B. 1987. Chesapeake Bay dissolved oxygen dynamics: roles of phytoplankton and microheterotrophs, pp. 75-84. In: G.B. Mackiernan (ed.) Dissolved Oxygen in the Chesapeake Bay: Processes and Effects. Publica-tion Number UM-SG-TS-87-03. Maryland Sea Grant College Program, College Park.

Jonas, R.B., D. Gluckman, J.H. Tuttle and J.T. Bell. 1988a. Distribution and metabolism of amino acids in Chesapeake Bay: Water column maxima. Presented at the American Geophysical Union, American Society for Limnology and Oceanography Ocean Sciences Meeting, New Orleans, Louisiana, January 1988.

Jonas, R.B., J.H. Tuttle, J.T. Bell and D.G. Cargo. 1988b. Organic carbon, oxygen consumption and bacterial metabolism in Chesapeake Bay. Presented at the U.S.E.P.A., Chesapeake Research Consortium conference, Understanding the Estuary: Advances in Chesapeake Bay Research, Baltimore, Maryland, March 1988.

Jonas, R.B., J.H. Tuttle, D.L. Stoner and H.W. Ducklow. 1988c. Dual-label radioisotope method for simultaneously measuring bacterial production and metabolism in natural waters. Appl. Environ. Microbiol. 54:791-798.

Jonas, R.B., J.H. Tuttle and S. Coon. 1989. The impact of oyster abundance on water quality in sub-estuaries of Chesapeake Bay, pp. 91-110. In: Research Proposal: Environmental Effects Research on Chesapeake Bay, Dissolved Oxygen and Related Processes. Maryland and Virginia Sea Grant College Programs.

Jonas, R.B. and J.H. Tuttle. 1990. Bacterioplankton and organic carbon dynamics in the lower mesohaline Chesapeake Bay. Appl. Environ. Microbiol. Vol 56.

Karl, D.M. 1982. Selected nucleic acid precursors in studies of aquatic microbial ecology. Appl. Environ. Microbiol. 44:891-491.

Kemp, W.M. and W. R. Boynton. 1980. Influence of biological and physical processes on dissolved oxygen dynamics in an estuarine system: Implications for measurement of community metabolism. Estuar. Coast. Mar. Sci. 11:407-431.

Kemp, W.M., W.R. Boynton and P. Sampou. 1987. Relative roles of benthic vs. pelagic processes in establishing and maintaining anoxia in Chesapeake Bay, pp. 103-114. In: G.B. Mackiernan (ed.) Dissolved Oxygen in the Chesapeake Bay: Processes and Effects. Publication Number UM-SG-TS-87-03. Maryland Sea Grant College Program, College Park.

Lagera, L.M.,Jr. and J.C. Zieman Jr. 1988. The relative significance of macrophyte decomposition and phytoplankton respiration in the consumption of oxygen in the lower Chesapeake Bay. Presented at the U.S.E.P.A., Chesapeake Research Consortium conference, Understanding the Estuary: Advances in Chesapeake Bay Research, Baltimore, Maryland, March 1988.

Lovell, C.R. and A. Konopka. 1985a. Primary and bacterial production in two dimictic Indiana lakes. Appl. Environ. Microbiol. 49:485-491.

Malone, T.C., W.M. Kemp, H.W. Ducklow, W.R. Boynton, J.H. Tuttle and R.B. Jonas. 1986. Lateral variation in the production and fate of phytoplankton in a partially stratified estuary. Mar. Ecol. Prog. Ser. 32:149-160.

McManus, G.B. 1989. Phytoplankton trophodynamics in Chesapeake Bay: Bacterivory and herbivory by nano- and micro-zooplankton, pp. 140-156. In: Research Proposal: Environmental effects research on Chesapeake Bay, Dissolved oxygen and related processes. Maryland and Virginia Sea Grant College Programs.

Newcombe, C.L. and W.A. Horne. 1938. Oxygen-poor waters of the Chesapeake Bay. Science 88:80-81.

Newell, R.I.E. 1988. Ecological changes in Chesapeake Bay: Are they the result of overharvesting the American oyster, *Crassostrea virginica?* In: Understanding the Estuary: Advances in Chesapeake Bay Research. Proceedings of a conference, March 1988. Publication 129. Chesapeake Research Consortium, Gloucester Point, Virginia.

Palumbo, A.V. and R.L. Ferguson. 1978. Distribution of suspended bacteria in the Newport River estuary, North Carolina. Estuarine Coastal Mar. Sci. 8:56-60.

Peele, E.R. 1988. Flux of carbon through bacterioplankton in Chesapeake Bay. Presented at the American Geophysical Union, American Society for Limnology and Oceanography Ocean Sciences Meeting, New Orleans, Louisiana, January 1988.

Robarts, R.D., R.J. Wicks and L.M. Sephton. 1986. Spatial and temporal variations in bacterial macromolecule labeling with [methyl-$^{3}$H]thymidine in a hypertrophic lake. Appl. Environ. Microbiol. 52:1368-1373.

Sale, J.W. and W.W. Skinner. 1917. The vertical distribution of dissolved oxygen and the precipitation by salt water in certain tidal areas. J. Franklin. Inst. 184(Dec.):837-848.

Sampou, P., W.T. Randall and W.M. Kemp. 1988. Planktonic respiration in mesohaline waters of Chesapeake Bay: Seasonal patterns of size-fractionated oxygen consumption with reference to depletion of bottom water oxygen. In: M.P. Lynch and E.C. Krome (eds.) Understanding the Estuary: Advances in Chesapeake Bay Research. Proceedings of a Conference, 29-31 March, Baltimore, Maryland. Publication Number 129, CBP/TRS 24/88. Chesapeake Research Consortium, Gloucester Point, Virginia.

Taft, J.L., W.R. Taylor, E.O. Hartwig and R. Loftus. 1980. Seasonal oxygen depletion in Chesapeake Bay. Estuaries. 3:242-247.

Tuttle, J.H., R.B. Jonas and T.C. Malone. 1987a. Origin, development and significance of Chesapeake Bay anoxia, pp.442-472. In: Majumdar, S.K., L.W. Hall, and H.M. Austin (eds.) Contaminant Problems and Management of Living Chesapeake Bay Resources. Pennsylvania Academy of Sciences Press, Easton, Pennsylvania.

Tuttle, J.H., T. Malone, R. Jonas, H. Ducklow, and D. Cargo. 1987b. Nutrient dissolved oxygen dynamics: Roles of phytoplankton and microheterotrophs under summer conditions,1985. CBP/TRS 3/87. U.S. Environmental Protection Agency-Chesapeake Bay Office.

Tuttle, J.H., E.E. Roden and C.L. Divan. 1987. Contribution of sulfur cycling to anoxia in Chesapeake Bay, pp. 100-102. In: G.B. Mackiernan (ed.) Dissolved oxygen in the Chesapeake Bay: Processes and Effects. Publication Number UM-SG-TS-87-03. Maryland Sea Grant College Program, College Park.

Wright, R.T. and R.B. Coffin. 1984. Factors affecting bacterioplankton density and productivity in salt marsh estuaries, pp. 485-494. In: C.A. Reddy and M.J. Klug (eds.) Current Perspectives in Microbial Ecology. American Society of Microbiology.

Zieman, J.C. and S.A. Macko. 1988. Stable isotope ratios of ecological components in South Florida and Chesapeake Bay: Comparisons of variation due to sources and processing. Presented at the American Geophysical Union, American Society for Limnology and Oceanography Ocean Sciences Meeting, New Orleans, Louisiana, January 1988.

# Benthic-Pelagic Interactions: Nutrient and Oxygen Dynamics

*W. Michael Kemp*
Horn Point Environmental Laboratory
Center for Environmental and Estuarine Studies
University of Maryland System, Cambridge, Maryland 21613

*Walter R. Boynton*
Chesapeake Biological Laboratory
Center for Environmental and Estuarine Studies
University of Maryland System, Solomons, Maryland 20688

## Introduction

In Chesapeake Bay primary production of organic matter tends to be dominated by phytoplankton; such production is autotrophic in that phytoplankton synthesize their own food from inorganic nutrients and sunlight. Though autotrophic activity occurs in the pelagic zone (throughout the water column), it is generally restricted to the euphotic, or upper, region of the water through which there is enough light penetration for photosynthesis to occur. Much of the consumption of organic matter, a heterotrophic process, is concentrated at or near the sediment surface (the benthos) where a great deal of biological and chemical activity occurs. The various mechanisms by which these pelagic and benthic zones are functionally connected have been referred to as pathways of *benthic-pelagic coupling*. It is the dynamic nature of this coupling that influences the levels of dissolved oxygen in the Chesapeake Bay estuary.

Two major pathways which have received much attention in recent years include (1) delivery of particulate organic matter (POM) from the water column, where it is produced, to the benthos, where it is stored and consumed; and (2) decomposition of particulate organic matter in the benthos and the resulting regeneration and transport of nutrients from sediments back to the euphotic zone, where they are assimilated again by phytoplankton for primary production. In some cases, benthic-pelagic coupling also provides a means by which organic and inorganic materials may be transported horizontally (across salinity and depth gradients) while being used reciprocally in autotrophic and heterotrophic processes (Kemp and Boynton 1984; Malone et al. 1986).[1]

The cumulative effect of complex pathways of benthic-pelagic coupling on dissolved oxygen concentrations in Chesapeake Bay and other estuaries depends largely on physical hydrodynamic processes (Boicourt, this volume). For example, density and pressure gradients, turbulence and internal waves set up by wind, tides and riverflow all affect the nature of interactions between benthic and pelagic components of the ecosystem. It is this acute dependence of ecological processes on physical transport which distinguishes aquatic systems, and particularly estuaries, from other major ecosystems.

Figure 1 depicts the principal processes of benthic-pelagic coupling for estuarine ecosystems. Nutrients — in particular, nitrogen, phosphorus and silicon — enter the estuary each year primarily during spring runoff/riverflow and support the new growth of phytoplankton communities. These communities are dominated by diatoms in spring and flagellated forms in summer (Malone, this volume). Much of this phytoplankton production of organic matter is

[1] On the one hand, the fate of many pollutants entering an estuary like Chesapeake Bay can be controlled by the nature of benthic-pelagic coupling mechanisms, while on the other, these benthic-pelagic interactions may themselves be altered by the introduction of various anthropogenic substances into coastal waters.

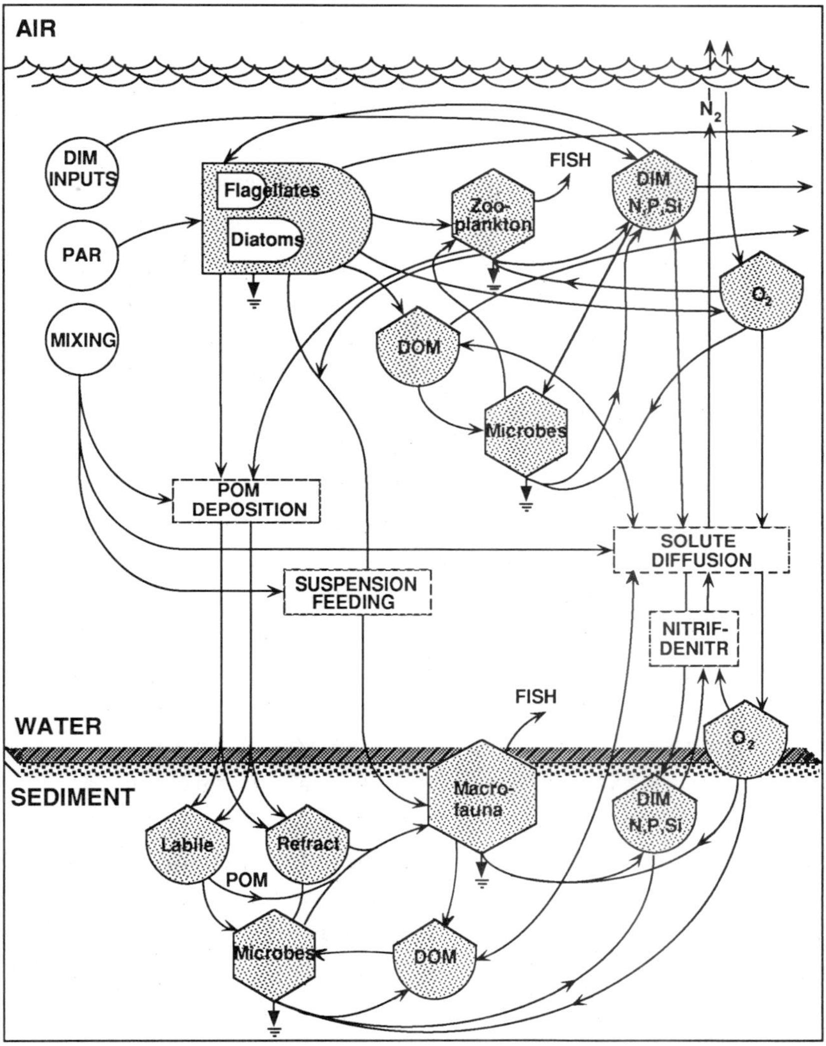


*Figure 1. Conceptual diagram depicting three major ecosystem processes of benthic-pelagic coupling in an estuarine environment: (1) deposition of particulate organic matter (POM); (2) diffusion of solutes across the sediment-water interface; (3) benthic macrofaunal suspension feeding on POM in overlying waters. Other processes shown include (1) plankton trophic interactions which affect POM deposition; (2) recycling of dissolved inorganic matter (DIM: N,P,Si) via planktonic and benthic metabolic processes; (3) the planktonic microbial loop mediated by excretion of dissolved organic matter (DOM); and (4) oxygen exchanges and effects on redox-sensitive processes such as nitrification-denitrification.*

consumed in the water column by zooplankton grazing and by microbial processes which involve dissolved organic matter excretion, bacterial consumption and protozoan grazing (Jonas, this volume). A significant portion of this production may also sink through the water column and be deposited as particulate organic matter to the sediment surface; in general this organic matter will be in the form of intact diatom cells or zooplankton fecal pellets.

The transfer of planktonic particulate organic matter to the sediments can be facilitated substantially by the suspension-feeding and active pumping of benthic macrofauna, such as clams and oysters. These invertebrates metabolize organic matter and convert it to biomass, which is in turn consumed by fish through a variety of pelagic and benthic food chains.

Protozoa, bacteria and metazoan animals — as a consequence of their metabolic consumption of the organic matter produced by phytoplankton — excrete dissolved inorganic nutrients. These recycled nutrients are then available to support further growth of phytoplankton.

Nutrient recycling in the planktonic subsystem tends to be rapid, while cycling in the benthos is slower: there are significant time delays between the large quantities of particulate organic matter deposition, decomposition and vertical diffusion (recycling) of nutrients back to the overlying water. For example, certain labile or rapidly degraded organic compounds such as sugars, lipids and amino acids in the particulate organic matter deposited to the sediment surface are decomposed readily; others such as lignins and cellulose are more refractory, or resistant to decomposition. Differences in decomposition of these compounds result in a spectrum of recycling rates from the benthos. Nutrients contained in the benthic subsystem, in contrast to the water column, are less susceptible to physical transport from the estuary to the continental shelf. Thus, deposition of particulate organic matter to the bottom and the resulting benthic processes of nutrient recycling represent a mechanism of retaining nutrients which enter the estuary in winter and spring long enough to support continued phytoplankton production in summer and fall (Kemp and Boynton 1984).

In the upper portion of the water column, dissolved oxygen is generated through phytoplankton photosynthesis. Oxygen is also exchanged by diffusion across the air-water interface, and consumed in heterotrophic metabolism of bacteria, animals and plants. In the benthic subsystem and lower portion of the water column, oxygen is consumed either directly or indirectly by most heterotrophic processes; oxygen can only be replenished, however, by vertical exchange with the upper layer of the water column. Under conditions of vertical density stratification that occur in late spring and summer, a strong stratified boundary, or pycnocline, develops between dense, salty lower waters and lighter, fresh, oxygenated upper waters. Stratification impedes physical exchange between these layers (Boicourt, this volume), thus preventing oxygen from reaching the lower waters and leading to the condition of oxygen depletion, or hypoxia (Kemp and Boynton 1980). While oxygen decline as a result of stratification is a naturally occurring phenomenon in many estuaries, including portions of Chesapeake Bay, there is widespread evidence that the process is accentuated by increasing inputs of nutrients from anthropogenic sources (Officer et al. 1984).

The trend of increasing anthropogenic nutrient enrichment, or eutrophication, of coastal waters can significantly affect estuarine ecosystems by several mechanisms which directly involve benthic-pelagic coupling. Typically, nutrient enrichment fertilizes phytoplankton production and abundance to such an extent that the phytoplankton, or algal, blooms will blanket large surface areas of the Bay. These algae sink to bottom waters, where they are decomposed by heterotrophic processes which consume oxygen (Malone, this volume). If bottom water remains oxygenated, higher algal biomass and production will lead to larger rates of particulate organic matter deposition, which may result in increased production of such benthic macrofauna as oysters (Grassle et al. 1985). Otherwise, in regions susceptible to depletion of oxygen in bottom waters, nutrient enrichment and high production will lead to decreased macrofauna (Cederwall and Elmgren 1980).

Nutrient fertilization can cause changes in the species composition and trophic structure of plankton communities, which in turn affect the rates, timing and quality of particulate organic matter deposition to the benthos (Smetacek 1984). In fact, it appears that the proportion of primary production delivered to the benthos actually decreases with increasing production along a eutrophication gradient (Oviatt et al. 1986). Relative rates of nutrient recycling from sediments tend to increase with fertilization; this is especially true for ammonium recycling because nitrification and denitrification — the reduction of nitrate to nitrogen gas — will be inhibited by oxygen limitation under eutrophic conditions. It is likely, therefore, that eutrophication results both in shifts between plankton and benthic food chains and in recycling pathways.

To better understand the role sediments and overlying waters play in the development of hypoxia and anoxia in Chesapeake Bay, it is important to characterize spatial and temporal trends for major benthic-pelagic processes, the factors controlling these processes, and the effects of eutrophication on them.

During the last five years there have been several active research programs focusing on various aspects of benthic-pelagic coupling in upper Chesapeake Bay (above the Potomac River mouth). A long term monitoring program (supported by the Maryland Department of Environment) was established in 1984 to measure fluxes of oxygen and nutrients across the sediment-water interface at ten permanent sampling stations (Figure 2) in the mainstem Bay (4 stations) and in three major tributaries (Patuxent, Potomac and Choptank rivers — two stations each). Data on phytoplankton production and benthic macrofauna abundance are being collected at these stations as part of this monitoring effort. Vertical arrays of fixed sediment traps have also been established at one of these stations (M4) in the mainstem Bay. The cross-Bay Chop-Pax transect of five stations intersecting the sediment trap site was created subsequently as part of a study of factors regulating seasonal oxygen depletion in the Chesapeake Bay (see Introduction). In 1987 the geographic scope of these studies was expanded to include sediment-water exchanges in the lower Bay and provide data needed for calibrating water quality models (supported by the U.S. EPA). In

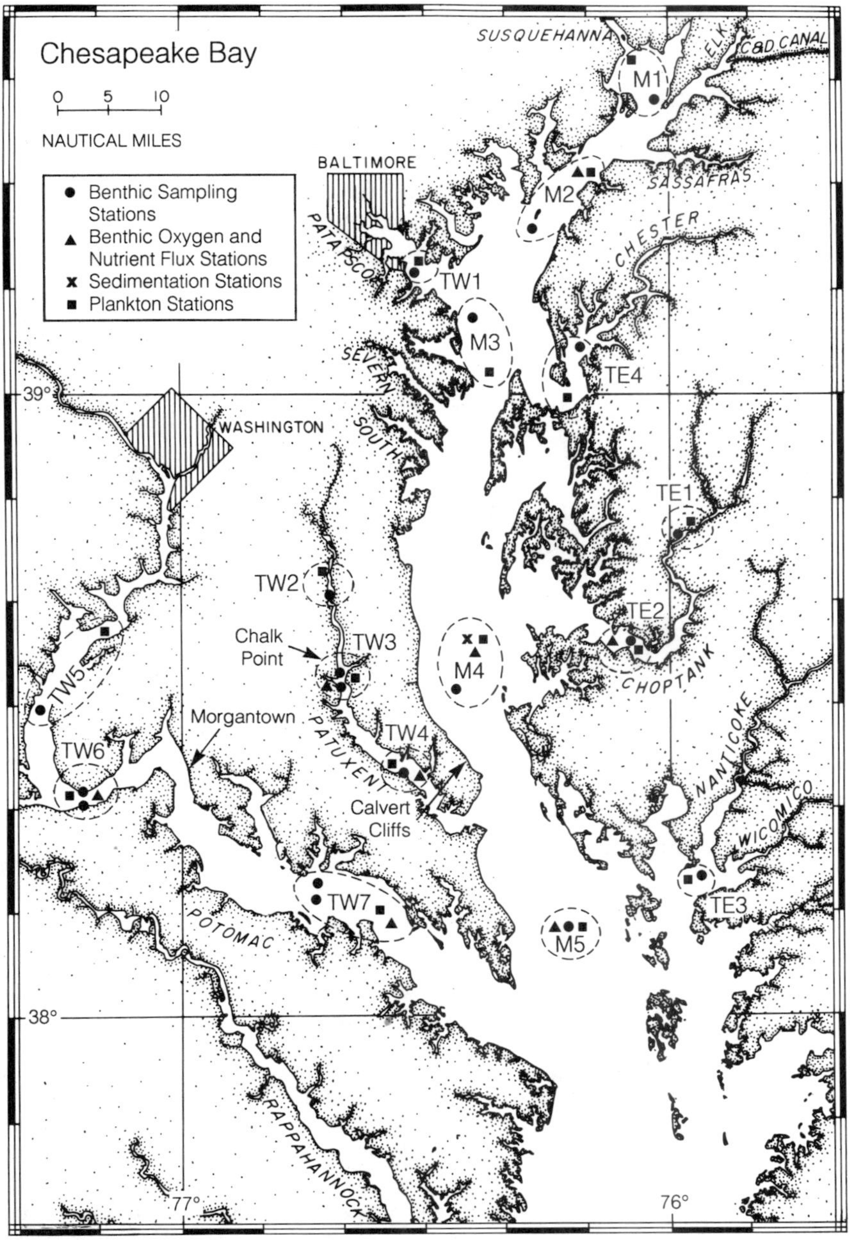


*Figure 2. Map of Maryland portion of Chesapeake Bay showing the location of sampling stations where benthic-pelagic processes are being measured in the Maryland Chesapeake Bay Water Quality Monitoring Program.*

addition, a large five-year research program was initiated in 1989 to investigate the fate and effects of nutrient inputs to estuaries as part of a new land-margin ecosystem research (LMER) initiative, supported by the National Science Foundation. Information generated from these diverse research and monitoring programs provides the primary basis for this chapter.

## Deposition of Particulate Organic Matter

One of the benthic-pelagic interactions most difficult to quantify is the deposition rates of particulate organic matter, largely because of problems in interpreting data from sediment traps.[2] Particle deposition rates have been measured using cylindrical traps that are deployed in the mesohaline reach of Chesapeake Bay. Seasonal patterns for particle deposition rates are relatively distinct, with three periods of high sedimentation generally occurring in April, August and October (Figure 3). These data, which have not been corrected for resuspension effects, represent particulate organic carbon (POC) and total chlorophyll *a* (Chl) collection rates in traps fixed at 9 m, which is the nominal depth of the pycnocline. Rates for 1985-1986 show that the broad seasonal trends appear to vary little from year

[2] In recent decades, there have been substantial developments in sediment trap methodologies which have improved our ability to measure particle deposition in lakes and oceans. Although the altered hydrodynamic flow fields induced by the traps themselves certainly affect particle dynamics around sediment traps (Gardner 1979a; Butman 1986; Butman et al. 1986), several design features can minimize these sampling biases. In accordance with recommendations of several trap analyses in flumes (Gardner 1979a,b; Bloesch and Burns 1980; Blomqvist and Hakanson 1981), sediment traps used in Chesapeake Bay (Boynton et al. 1988) have employed a cylindrical shape with a 10:1 ratio of height to diameter. In shallow, hydrodynamically active environments such as estuaries, resuspension (followed by sinking) of bottom sediments (Oviatt and Nixon 1975) tends to confound measurements of deposition of autochthonous, newly formed particles (e.g., algal cells and zooplankton fecal pellets). Sediment trap rates of POM settling can be corrected for contributions of resuspended material using a pro rata scheme which compares the % organic content of trapped material with those of bottom sediments and suspended material in the surface waters (Gasith 1975; Taguchi 1982). In addition, rates of total chlorophyll pigment

to year. Similar interannual consistency of seasonal patterns has also been reported for the Kiel Bight region of the southern Baltic Sea (Smetacek 1984). The spring and summer deposition events are closely associated with the annual maxima in euphotic zone chlorophyll *a* concentration and phytoplankton production, respectively. There is a small increase in chlorophyll *a* stocks during the summer period of higher deposition, but not so in the autumn. The seasonal trends are more pronounced for sedimentation of chlorophyll *a* than particulate organic carbon.

The carbon to chlorophyll ratio (C:Chl) of sedimented particulates varies seasonally in relation to the changing character of deposited material (Figure 3). The ratio is lower in the spring than in the summer, possibly indicating a shift in trapped material from intact algal cells to zooplankton fecal pellets, respectively (Steele and Baird 1972; Bodungen et al. 1981; Forsskahl et al. 1982; Smetacek 1984). Indeed, microscopic examination of these materials revealed that in 1986, 97% of the total mass trapped in April was associated with centric diatoms, while 76% in August was fecal pellets (K. Sellner, personal communication). The rapid deposition of diatoms from April to May is consistent with typical life-cycle sequences for diatoms (Smetacek 1985) and may result from incipient silicon limitation for their growth. The dominance of fecal pellets in the trapped

---

deposition estimated with sediment traps are less susceptible to resuspension effects because of the relatively rapid rate of pigment decomposition in bottom sediments (e.g., Carpenter et al. 1986).

Another potential problem with interpreting sediment trap data is the decomposition of organic particles within the traps occurring during the deployment period (Honjo and Roman 1978; Ducklow et al. 1985). In attempting to rectify this problem, some investigators have added preservatives to the traps to retard bacterial activities (Knauer et al. 1984); however, others have observed that preservatives can further complicate interpretations by causing an accumulation of dead invertebrates which swim into the trap (Peterson and Dam 1990). Experiments have been conducted to test effectiveness of preservatives for sediment traps in Chesapeake Bay. Although Chl rates were unaffected by poisoning, particulate organic carbon was increased significantly, suggesting artificially induced mortality of "swimmers" (Boynton et al., unpublished). In general, the relatively short deployment intervals (4 to 7 days) used in the Bay sediment trap studies appear to be sufficient to minimize problems associated with degradation of particulate organic matter in traps (Boynton et al. 1988).

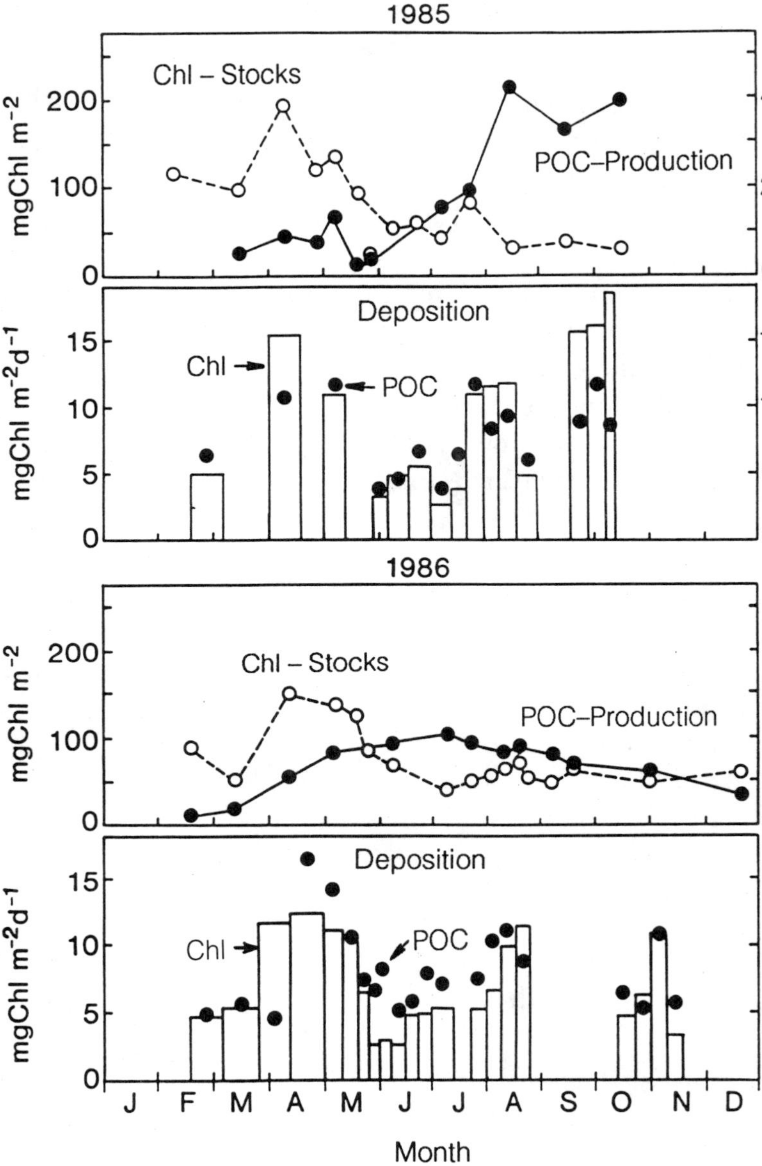


*Figure 3. Annual patterns of chlorophyll* a *(Chl) stocks and deposition rates and particulate organic carbon (POC) production and deposition rates in 1985 and 1986 in the mesohaline region of Chesapeake Bay (Boynton et al. 1988).*

material in summer is coincident with the timing of maximal zooplankton grazing (Malone, this volume).

In addition, the relative rate at which euphotic zone chlorophyll *a* biomass is depleted via sinking can be inferred from the slope of chlorophyll *a* deposition versus chlorophyll *a* stocks (Billen and Lancellot 1988). Using data in Figure 3 from both years, statistically significant relationships between chlorophyll *a* deposition and concentration are obtained for both spring ($r^2 = 0.88$) and summer ($r^2 = 0.70$), with the slope for the latter period being twice that of the former. Evidently, the phytoplankton community turns over via sinking at a rate of approximately once every 14 days in spring and once every 7 days in summer. In spring, the relatively ungrazed algal stocks which accumulate with great abundance are dominated by large diatoms, and even though sinking rates are high, the algal biomass turns over more slowly. In contrast, the smaller flagellated cells which predominate in summer are heavily grazed with less build-up of biomass, so that turnover of algal stocks is faster even though rates of particulate organic matter sinking are lower. From the perspective of the benthic community, the food quality of intact diatom cells deposited in the spring probably exceeds that of the fecal material sedimenting in summer (Marsh and Tenore 1990).

Although much of the particulate organic matter caught in sediment traps derives from the sinking of algal cells and fecal pellets that are produced in the overlying euphotic zone, indirect evidence suggests that a considerable fraction is from production occurring in the shallower regions flanking the main Bay channel. For example, significant correlations were observed between phytoplankton production in both the east and west flanks and sediment trap collections in the main channel; however, no relation was seen between deposition and production in the overlying water column (Malone et al. 1986). Sediment traps deployed 1 m above the sediment surface (20 m deep) in the channel collected 5 to 10 times more particulate organic matter than in the pycnocline and euphotic zone traps described above (Boynton et al. 1988), even though most of the resuspension of bottom sediments occurs at depths less than 10 m (Ward 1985). This observation further emphasizes the importance of lateral transport of resuspended particulate material from shallow waters flanking the Bay channel to the channel bottom.

This process of bottom resuspension and lateral transport to pycnocline traps could cause serious overestimates of deposition of newly produced (as opposed to resuspended) particulate organic matter from overlying water. However, such effects would be expected to be minimal for chlorophyll *a*, because these algal pigments degrade relatively rapidly on the sediment surface and are thus less available for resuspension. The seasonal patterns of chlorophyll *a* deposition estimated with traps and reported in Figure 3, therefore, are probably representative of actual conditions (see footnote 2). The effects of resuspension on estimates of new particulate organic carbon deposition can be corrected by the ratio of new:total carbon deposited estimated by pro rata of the percent of organic carbon in seston, bottom sediments and trapped particulates ( = $[\%C_{seston} - \%C_{bottom}]\,[\%C_{trap} - \%C_{bottom}]^{-1}$).

Annual rates of new particulate organic carbon deposition were estimated for 1985 at the mesohaline Bay trap by applying the above corrections (factors ranged from 0.4-0.9 over the year), and these rates compared favorably (within 20%) to particulate organic carbon deposition calculated from annual chlorophyll *a* rates, multiplied by the C:Chl ratio for surface particulates (Boynton et al. 1988). In addition, annual rates for deposition of total dry mass, estimated from the particulate organic carbon rates divided by the %C of surface seston, minus the loss of mass associated with sediment respiration, were compared to long-term sediment burial rates obtained from $^{210}Pb$ analyses (Officer et al. 1984). Mean rates for 1985 sediment traps were 0.24 g dry wt $cm^{-2}y^{-1}$ compared to $^{210}Pb$ rates which ranged from 0.1-0.3 g dry wt $cm^{-2}y^{-1}$. While these annual rates have been calculated for comparative purposes and must be considered crude estimates, they do lend credibility to the quantitative aspects of the Chesapeake Bay sediment trap data. Similar close correspondence between sediment trap and $^{210}Pb$ estimates of deposition have been reported in lacustrine environments (Bloesch and Evans 1982).

A preliminary calculation for the balance of organic carbon metabolism in the water column at the mesohaline site of the sediment traps provides yet another test of trap rates with regard to actual particulate organic carbon deposition. In this analysis, all rates are based on oxygen measurements converted to carbon, assuming photosynthetic and respiratory quotients of 1.2. Previous measurements (Kemp and Boynton 1980; 1981) at a nearby site (6 m depth) revealed significant correlations between plankton production and both the respiration of the plankton ($r^2 = 0.66$) and of the benthic communities ($r^2 = 0.35$). These measurements suggest that production and respiration are closely coupled, and the combined slopes of these relations indicate that 85-90% of the autochthonous production — primary production that occurs in the water column — is consumed in place.

On time scales of days to weeks, however, particulate organic carbon deposition is poorly correlated with primary production, and it is the variability of the respiration-production relationship which may be more important than production alone in determining deposition rates. Using 1986 measurements from the sediment trap station and subtracting the upper layer (0-8 m) and lower layer (8-20 m) water column respiration rates from this production, a "residual" term is left in the carbon budget (Figure 4) that corresponds to the particulate organic carbon available for deposition to the sediment surface. If this residual, calculated at 2-4 week intervals, is compared to sediment trap collection rates for the same time periods, a remarkably close correlation is obtained (slope=0.94, $r^2$=0.96), further supporting the quantitative robustness of the traps.

### *Comparisons with Other Systems*

Measurements of various aspects of benthic-pelagic coupling throughout the world permit comparisons of results with Chesapeake Bay. For example, strong correlations are evident in a variety of systems between planktonic primary production (adjusted for depth of the upper mixed layer) and annual sedimentation of organic matter (Figure 5a) (Suess 1980; Hargrave 1984). Annual data from studies in Chesapeake Bay fit the general pattern and are at the top end of data from shallow systems. In a similar fashion the percentage of annual primary production that is sedimented appears to be a function of depth (Suess 1980), ranging from as much as 80% in shallow systems to less than 1% in the deep ocean. In systems less than 50 m in depth (Figure 5), proportionally less organic material sediments from the water column per unit of production than in deeper systems; this is an indication of a shift to more thorough utilization of organic material in the water column. Again, data from Chesapeake Bay fit this pattern: between 20-80% of annual production is sedimented.

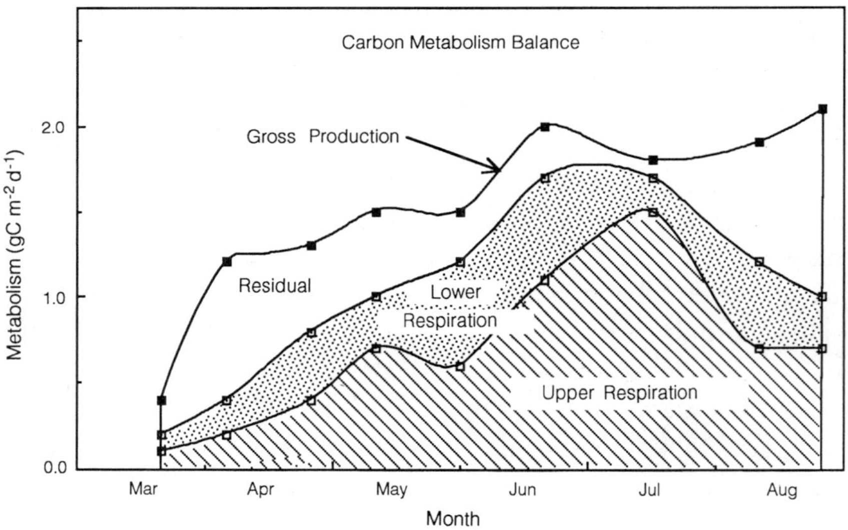


*Figure 4. Calculated seasonal balance of: organic carbon gross production, plankton respiration in the upper layer, and plankton respiration in the lower layer. The residual is taken as production minus upper and lower layer respiration rates (Kemp et al. unpublished).*

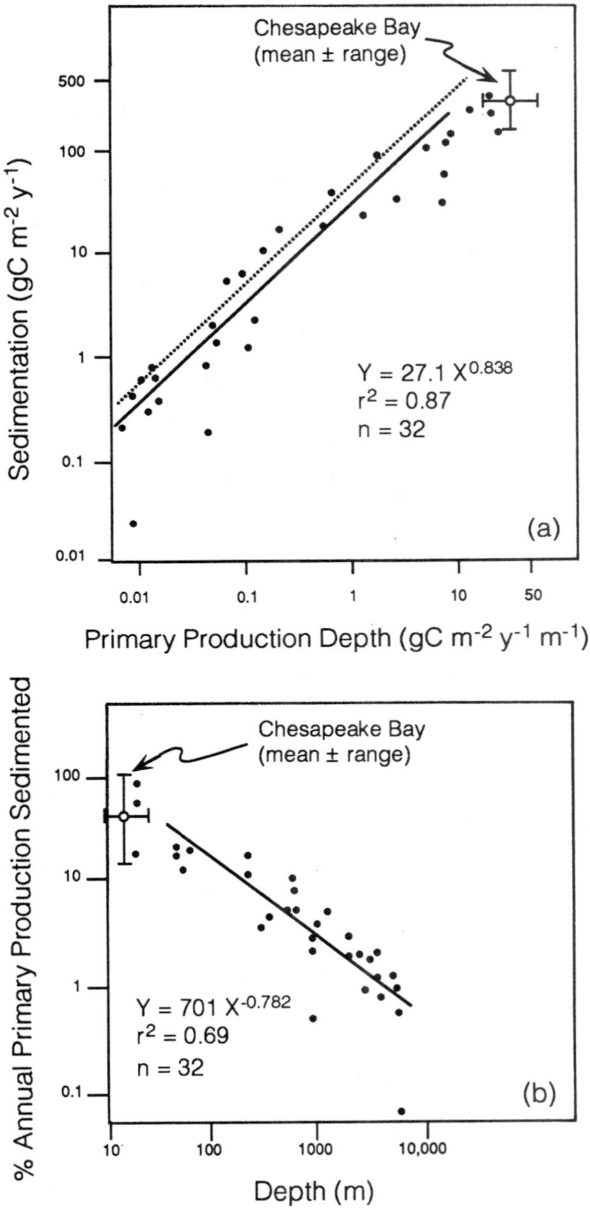


*Figure 5. Summary of benthic-pelagic coupling processes from the literature (Hargrave 1984) with values for annual means and ranges from Chesapeake Bay (Boynton et al. 1988; Holland 1988; Sellner et al. 1988) presented for comparisons: (a) sedimentation versus production per mixed layer depth; (b) % production sedimented versus water column depth.*

## Benthic Metabolism and Nutrient Cycling

Exchanges of solutes across the sediment-water interface have been measured routinely in numerous coastal marine environments (Harrison 1980; Nixon 1981; Nixon and Pilson 1983; Klump and Martens 1983). Typically, these rates are estimated from changes in solute concentrations in water enclosed under chambers placed on the sediment surface or in water overlying intact sediment cores incubated in temperature-controlled laboratories. In principle, total metabolism of the benthic community can be estimated as uptake of oxygen or as efflux of dissolved inorganic carbon ($TCO_2$). Burial and temporary storage of the end-products of anaerobic respiration (primarily sulfide) can confound interpretations of oxygen flux as community metabolism. Although only a small fraction (5 to 15%) of the sulfide formed in sulfate reduction is buried permanently, it is thought that significant quantities are temporarily stored as iron-sulfides (Jørgensen 1983). This storage implies that, while instantaneous rates of oxygen exchange and community metabolism may not correspond to one another, integrated annual rates may be comparable. Indeed, recent analyses indicate that annual rates of sediment oxygen consumption (SOC), sulfate reduction and dissolved inorganic carbon efflux compare favorably for the mesohaline region of Chesapeake Bay (E. Roden, personal communication).

### *Sediment Water Fluxes*[3]

Seasonal sequences for sediment-water fluxes of dissolved oxygen ($O_2$), ammonium ($NH_4^+$), nitrate ($NO_3^-$), silicon (Si) and phosphate ($PO_4^{3-}$) are available for ten stations in the upper and middle Bay and for three major tributaries since 1985 (Boynton et al. 1988). Mean patterns for the upper and middle main Bay stations are summarized in Figure 6; they illustrate representative trends in the

[3] The following discussion focuses on information available since 1987 with regard to seasonal and spatial patterns of solute fluxes across the sediment surface in upper Chesapeake Bay and its tributaries. Garber (1988) thoroughly reviewed data until 1986.

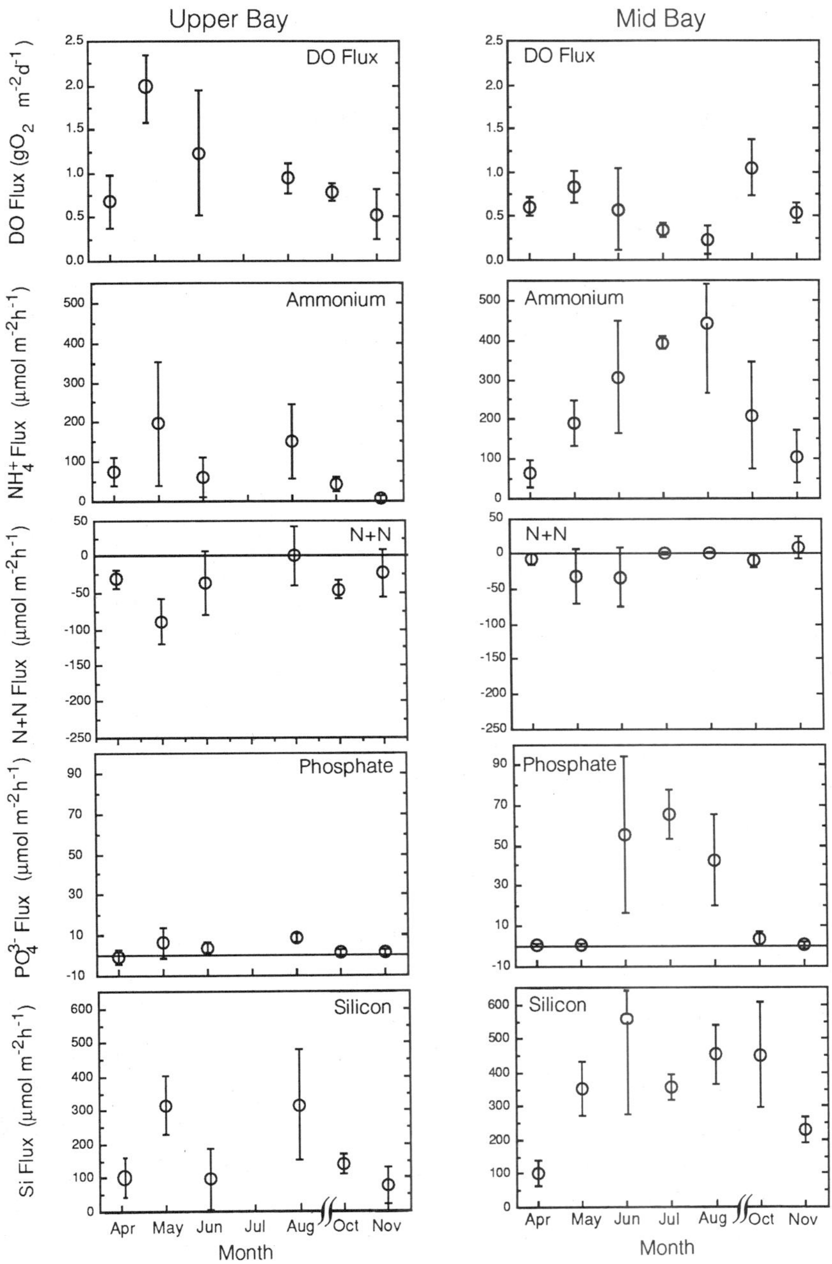


*Figure 6. Composite of annual cycles (mean ± S.E.) for sediment-water exchanges of oxygen, ammonium, nitrate plus nitrite (N+N), silicate and phosphate for 1984-1988 at two stations (Sta. M2 and M4, Figure 2) in the main stem of Chesapeake Bay (Boynton et al. 1988).*

estuary. Seasonal maxima for dissolved oxygen flux occurred in May for the upper Bay site and in October for the mesohaline Bay, with slightly higher mean annual rates in the upper Bay. The low rates in August for the mid Bay were related to the dissolved oxygen depletion which often prevails at this time. Under anoxia there was, obviously, no dissolved oxygen flux at the sediment surface, but the sulfide diffusing from sediments to anoxic overlying water consumed dissolved oxygen at the pycnocline.

Rates of ammonium efflux from sediments for both stations tended to be highest in August, but annual mean values were 2 to 3 times higher at the mid-Bay site. This seasonal pattern closely follows that for water temperature. In contrast, fluxes of nitrate (plus nitrite, $NO_2^-$) were generally directed into the sediments. However, the seasonal patterns differed considerably between the upper- and mid-Bay sites: at the upper-Bay site, seasonal mean fluxes ranged from about −80 to −10 μmol $m^{-2}h^{-1}$ with no apparent relationship to nitrate concentrations in overlying waters. At the mid-Bay site, maximum fluxes were about half those observed at the upper Bay but were clearly related to the pattern of nitrate in overlying waters with highest fluxes being recorded in the spring and then declining through the summer period as the nitrate associated with the spring freshet was depleted. While the general pattern of nitrate flux at both stations was from water to sediments, there are times of the year when nitrate released from these sediments is indicative of nitrification activity in excess of dentrification.

At the upper-Bay site nitrate release has been frequently observed during summer and fall when nitrate concentrations in overlying waters are reduced although the magnitude of these fluxes is not sufficient to change the 1984 to 1988 average from negative to positive. Positive nitrate fluxes at the mid-Bay site are more sporadic and seemingly occur whenever the sediments are in contact with oxygenated waters and $NO_3^-$ concentrations in overlying waters are relatively low (Boynton and Kemp 1985). These trends in sediment-water fluxes of dissolved inorganic nitrogen are similar to those reported previously for Bay sites (Boynton et al. 1980; Callender and Hammond 1982; Boynton and Kemp 1985).[4]

Fluxes of silicon exhibited a bimodal pattern at the upper Bay site, ranging from about 300 $\mu$mol $m^{-2}h^{-1}$ in the early spring and summer to around 100 $\mu$mol $m^{-2}h^{-1}$ in the late spring and fall. In contrast to this complex pattern, fluxes of silicon at the mid-Bay site exhibited a strong unimodal pattern with highest fluxes observed following the completion of the spring diatom bloom but prior to the annual temperature maxima. Fluxes of silicon at the mid-Bay site are also among the highest observed throughout the Bay system with some measurements in excess of 1000 $\mu$mol $m^{-2}h^{-1}$. Sediment releases of silicon at locations 40 km up and downstream of the mid-Bay site were significantly smaller, suggesting that the mid Bay region may be a focal point of diatom deposition fueling the substantial releases observed.

Net exchanges of phosphate were distinctively different between sites (Figure 6). At the low salinity upper Bay site phosphate fluxes were always relatively low (< 15 $\mu$mol $m^{-2}h^{-1}$) and occasionally were directed into sediments. Average annual values were about 4 $\mu$mol $m^{-2}$ $h^{-1}$. Fluxes of this magnitude have little influence on ambient concentrations of phosphate in overlying waters. Fluxes of phosphate at the mid-Bay site were very large during the summer period, averaging about 50 $\mu$mol $m^{-2}h^{-1}$ and ranging as high as 120 $\mu$mol $m^{-2}h^{-1}$. At this site fluxes were low whenever bottom water oxygen concentrations were above 1 to 2 mg $L^{-1}$. It is apparent that redox conditions of sediments exerted a strong influence on these phosphate fluxes.

With the exception of oxygen fluxes, sediment-water exchanges at the mid-Bay site were considerably larger than those at the upper-Bay location and were, without exception, characterized by strong

[4] Similar seasonal patterns of ammonium fluxes, with peak rates in mid summer, have been described for a wide variety of coastal environments. However, in contrast to the patterns in the Bay, nitrate sediment-water exchanges exhibited a predominance of effluxes from sediments throughout the year for most other sites (Davies 1975; Nixon 1981; Fisher et al. 1982; Hopkinson and Wetzel 1982; Flint 1985; Hopkinson 1987).

unimodal seasonal patterns with peak rates in the summer. The magnitudes and seasonal patterns of oxygen, silicon and phosphate fluxes are within the ranges of those reported previously for a diversity of coastal marine environments (Nixon 1981; Fisher et al. 1982; Hammond et al. 1985; Hopkinson 1987).

In addition to the distinctive seasonal patterns of sediment-water fluxes, strong spatial variability was also evident. Figure 7 characterizes fluxes of oxygen, nitrate, phosphate, ammonium and silicate at 10 stations along the Bay's longitudinal axis for April and August. August fluxes of phosphate and ammonium, for example, were low in the Bay's upper and lower region, but reached maxima in the mid region. These high summer fluxes (Figure 3) were probably supported in part by the large amounts of organic matter reaching the sediment surface from direct deposition of locally produced organic matter and by horizontal transport of organic matter produced along the flanks of the Bay.

The largest fluxes occurred in the summer (August) when dissolved oxygen concentrations are typically low ($< 1$ mg $L^{-1}$). Some variability is evident in the spatial patterns of nitrate fluxes, but nitrate fluxes were highest in the upper Bay and decreased seaward. However, this pattern shifted dramatically between seasons with fluxes directed into sediments during April — presumably because of high nitrate concentrations in overlying waters — and out of the sediments to overlying waters in summer as a result of nitrification exceeding denitrification. Except in the upper Bay, nitrate fluxes were small in comparison to ammonium fluxes.

Strong seasonal differences in the magnitude and spatial pattern of fluxes for other nutrients were also observed. For example, ammonium and phosphate fluxes were quite small in April throughout the Bay, but were large and exhibited mid-Bay maxima in August. Sediment oxygen consumption, or oxygen fluxes, were rather uniform through the Bay in April (approximately $-0.7$ $gO_2$ $m^{-2}d^{-1}$), but tended to be higher at the freshwater and ocean ends and depressed in the central Bay regions during summer. Fluxes of silicon exhibited similar patterns in spring and summer, differing mainly in magnitude with summer fluxes being 2 to 5 times larger.

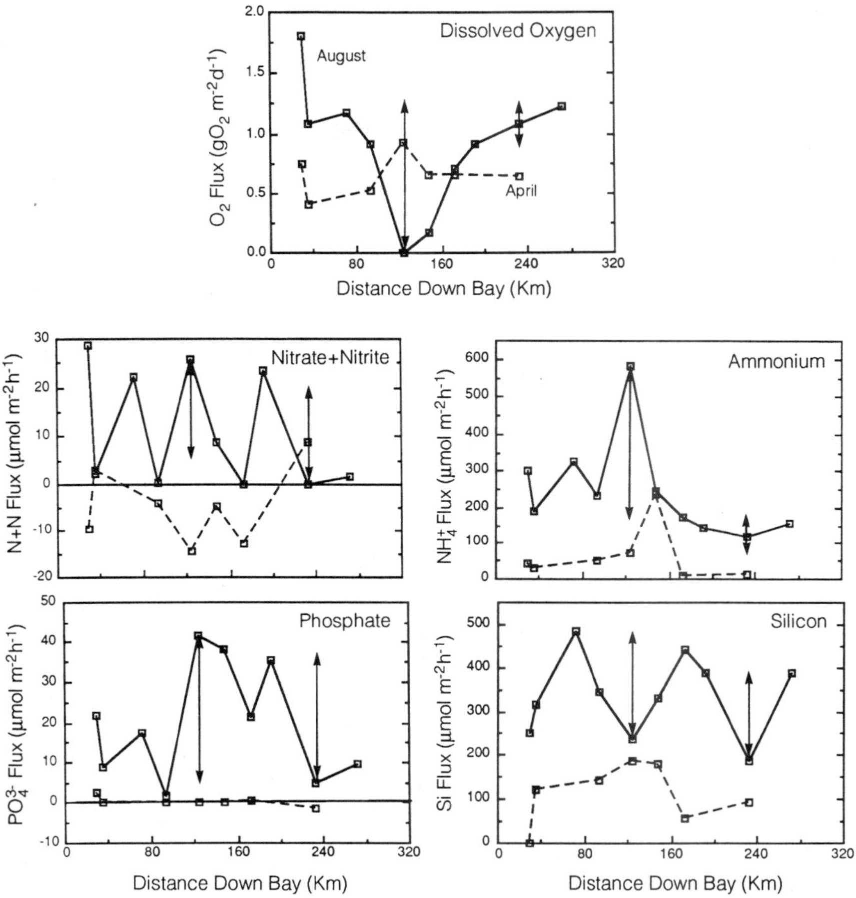


*Figure 7. Sediment-water exchanges of oxygen, nitrate plus nitrite (N + N), ammonium, phosphate and silicon for April and August 1988 at stations along the Chesapeake Bay mainstem, from the Susquehanna Flats to Cape Charles. Vertical arrows indicate range of rates for three stations in a cross-Bay transect measured in August in mid- and lower-Bay areas (Boynton et al. unpublished).*

During August 1988, two additional lateral stations were sampled in the mid and lower Bay to provide an indication of cross-bay variation in the magnitude of fluxes. As indicated in Figure 7, fluxes of some nutrient species differed considerably between shallower flank and deep stations. Some of these differences can be explained by higher oxygen concentrations in overlying waters and surficial sediments at the lateral stations. Lateral station differences at the lower Bay transect were not as pronounced nor were the differences in depth and redox conditions of overlying waters and sediments.

### *Environmental Factors*

A number of environmental factors are potentially important in regulating sediment-water fluxes in estuarine systems — among them temperature and redox conditions. In the mid-Bay region oxygen and ammonium fluxes follow the seasonal temperature cycle. Previous studies in the Patuxent River also found strong correlations between some nutrient fluxes and temperature (Boynton et al. 1980). When all data from all stations in the Bay since 1985 were combined, a strong relation was evident between oxygen concentration in overlying waters and phosphate fluxes. In this case, when oxygen concentrations in overlying waters were greater than about 1.5 mg $L^{-1}$, phosphate fluxes were small (being directed either into or out of sediments); only when oxygen concentrations were below 1.5 mg $L^{-1}$ were phosphate fluxes large, reaching 87 $\mu$mol $m^{-2}$ $h^{-1}$. Rates of sediment oxygen consumption were also correlated (though weakly) with oxygen concentrations in the overlying water (Boynton et al. 1988). The fact that there were also low phosphate fluxes observed under low oxygen conditions indicates that other environmental variables (e.g., quality and quantity of organic matter reaching the sediment surface, macrofaunal composition and abundance) are also involved in regulating phosphate fluxes.

One measure of the overall strength of benthic-pelagic coupling can be found in correlations between water column and sediment processes. For selected sites in the Bay where contemporaneous rates were available, annual estimates of sediment oxygen consumption, or oxygen fluxes, are plotted as a function of primary production in the water column (Figure 8). In this case, annual rates of primary production have been divided by the depth of the mixed layer to account for the potential losses of organic material to water column respiration (Hargrave 1973). As indicated in Figure 8a, there is a significant positive relationship between these processes, which explains about 50% of the observed variability. However, such simple relationships do not appear to exist for other sediment fluxes such as ammonium (Figure 8b). As was the case for phosphate fluxes, it appears that oxygen conditions in deep waters and surficial sediments strongly modify what might otherwise be a clear relationship indicative of direct benthic-pelagic coupling. At some sites with oxic overlying water, rates of ammonium releases may have been lower than anticipated from production in overlying waters, because of nitrogen losses associated with coupled nitrification-denitrification. On the other hand, relatively high ammonium releases under a range of organic matter loading regimes could result if the nitrification pathway was inhibited by low oxygen conditions (Kemp et al. 1990).

While deposition of autochthonous phytoplankton particulate organic matter to the sediments appears to influence annual rates of benthic metabolism (e.g., Figure 8), nutrient regeneration from benthic processes might, in turn, exert reciprocal influence on plankton production in overlying waters. Previous studies in Chesapeake Bay and its tributary estuaries have shown that rates of ammonium and phosphate release from sediments to overlying waters are quantitatively sufficient to support 20-80% of the demands for phytoplankton growth (Callender and Hammond 1982; Kemp and Boynton 1984; Boynton and Kemp 1985). In general, benthic recycling of both nitrogen and phosphorus represents larger fractions of respective plankton nutrient requirements in summer than in spring.

A conceptual model proposed by Kemp and Boynton (1984) suggests that plankton production is supported primarily by inputs of new nutrients from land in spring and benthic recycling in summer. They further postulate that the sequential processes — dis-

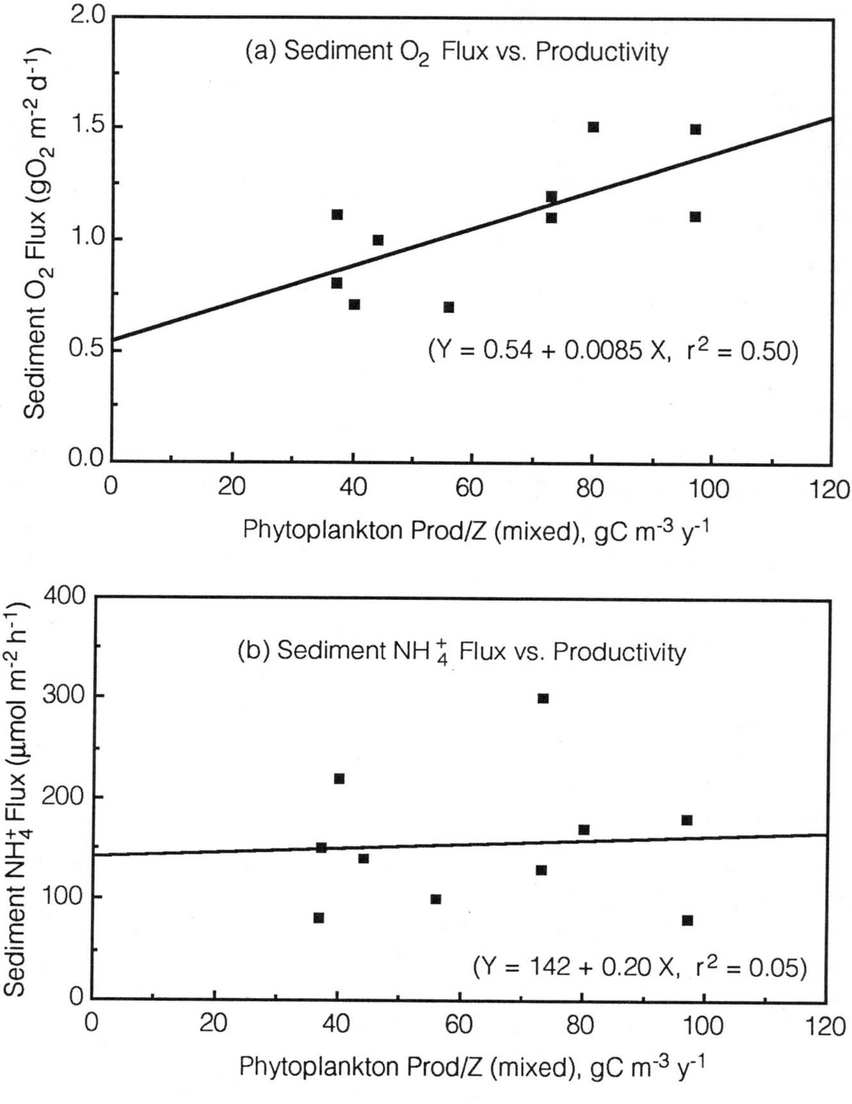


*Figure 8. Correlations between annual mean phytoplankton production divided by the mean depth of the upper mixed water column layer versus (a) annual mean rates of sediment oxygen consumption (SOC) and (b) annual mean sediment-water exchanges of ammonium. Each point represents a different station in the main stem or tributaries of the Bay (data taken from Boynton et al. 1988 and Sellner et al. 1988).*

solved nutrient uptake by phytoplankton, particulate organic matter deposition to sediments, benthic decomposition of particulate organic matter, and nutrient recycling to overlying water — present a means for retaining the winter-spring peak of nutrient inputs such that they can be used by phytoplankton in the summer peak of primary production.

A simple numerical simulation model of the major ecological processes involved in organic production and nutrient recycling was used to test this reasoning (Kemp and Randall 1988). Figure 9 summarizes simulation results for phytoplankton abundance (mgC $L^{-1}$) in the Bay's mesohaline region under two hypothetical scenarios, compared to the nominal base conditions (1986) for which the model was calibrated. In Figure 9a, nutrient inputs from the land were set to zero, resulting in virtual loss of the spring bloom, which is evidently supported primarily by "new nutrients" from external sources. In Figure 9b, vertical exchanges of nutrients from deep to euphotic waters were restricted in the model; this had the effect of greatly reducing summer phytoplankton production, which evidently depends on benthic recycling processes. Thus, the distinct seasonality of this benthic-pelagic coupling appears to play a strong role in regulating the magnitude and timing of summer production (Malone, this volume).

### *Comparisons with Other Systems*

Data are also available to compare plankton production and/or deposition with benthic metabolic processes and fluxes among a number of coastal ecosystems. Nixon (1981) found direct evidence of coupling between sediments and overlying waters by comparing benthic remineralization of carbon — actually, sediment oxygen consumption — with organic matter supply for a selection of shallow estuarine and coastal systems (Figure 10a). Data from Chesapeake Bay generally fall within the range observed for other systems, where a large fraction, approximately 50%, of organic inputs to the system (primary production plus import of allochthonous material) are metabolized in sediments.

The sites compared in Figure 10a are all relatively shallow

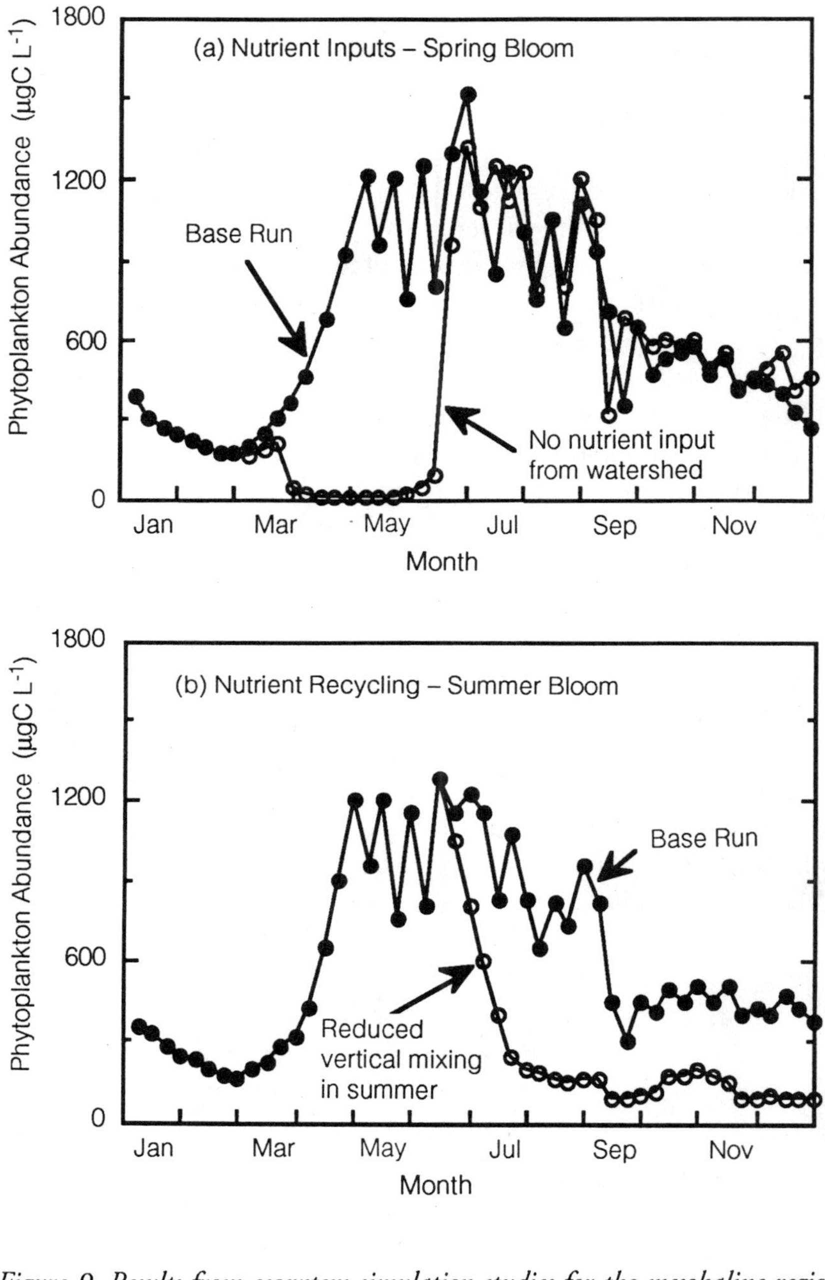


*Figure 9. Results from ecosystem simulation studies for the mesohaline region of Chesapeake Bay comparing nominal base-run conditions (1985) with (a) conditions with no new nutrient inputs from the watershed and (b) conditions with greatly reduced vertical mixing between bottom and upper layers of the water column (Kemp and Randall 1988).*

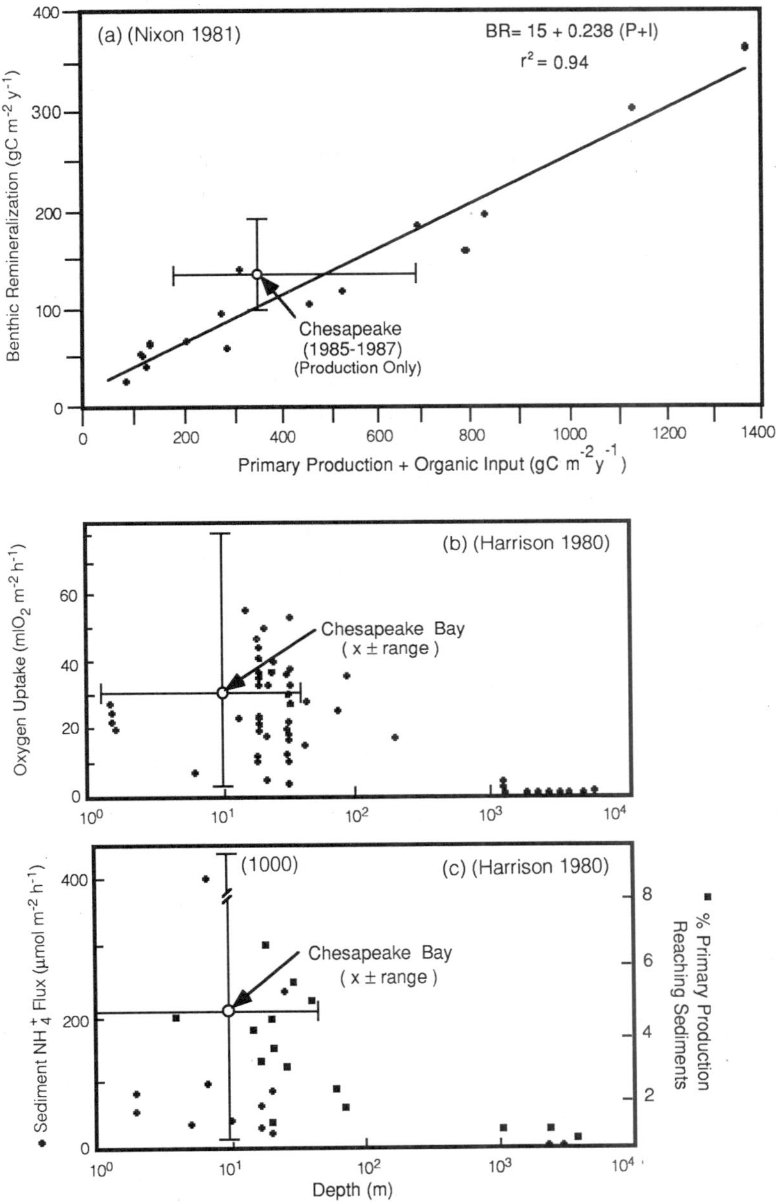


*Figure 10. Summary of benthic-pelagic coupling processes from the literature (Nixon 1981; Harrison 1980) with values for annual means and ranges from Chesapeake Bay (Boynton et al. 1988; Holland 1988; Sellner et al. 1988) presented for comparisons: (a) benthic remineralization versus primary production plus organic inputs; (b) benthic oxygen uptake versus depth; and (c) ammonium regeneration versus depth.*

(< 50 m), such that depth differences had little effect on the relation. Across a wider depth gradient, however, Harrison (1980) illustrated that sediment oxygen consumption rates decrease with water depth, presumably indicating a decrease in the fraction of organic production reaching sediments. There is considerable scatter in these data (Figure 10b) with the exception of deep ocean sites where sediment oxygen consumption rates were very low, as might be expected. Over the limited depth range of the Chesapeake Bay sites, sediment oxygen consumption values varied from close to zero to 80 $mlO_2\ m^{-2}h^{-1}$, exceeding the range of values observed from a variety of other systems. The highest rates for Bay stations correspond to regions of high particulate organic matter deposition. The low sediment oxygen consumption rates may, however, be related to low oxygen at some deep water stations in summer.

Harrison (1980) also compared sediment ammonium fluxes across a large depth range (Figure 10). Again, low fluxes were observed from sites located in waters in excess of 1000 m. In shallower systems ammonium fluxes were rather uniform with most being less than 100 $\mu mol\ m^{-2}h^{-1}$. Average fluxes from Chesapeake Bay exceed all but two observations from other systems and maximum values (approximately 1000 $\mu mol\ m^{-2}h^{-1}$) are among the highest yet recorded. While substantial ammonium fluxes have been found in many areas of the Bay's mainstem and tributaries (Boynton et al. 1988), highest fluxes have consistently been associated with areas exposed to low redox conditions. In these areas it appears that most of the particulate organic nitrogen deposited on the sediment surface is decomposed to form ammonium, which is recycled back to the water column rather than being temporarily sequestered in macrofaunal biomass or lost to nitrogen gas via coupled nitrification-denitrification.

## Sediment Nitrogen Cycling

Substantial evidence exists that nitrogen is the most important nutrient limiting growth and biomass accumulation of phytoplankton in Chesapeake Bay and other coastal waters (Boynton et al. 1982; D'Elia et al. 1986; Malone et al. 1988; Malone, this volume). Although phosphorus may also be an important limiting nutrient for algal growth in spring in oligohaline, or fresher water, sites and in general on geological time scales (Smith et al. 1987), nitrogen limitation tends to dominate on seasonal and annual scales in most estuaries. A number of explanations have been postulated for this pattern of nitrogen limitation in coastal marine systems. One such explanation is that high rates of loss of fixed nitrogen via denitrification in these estuarine environments results in nitrogen deficiency (Nixon 1981). Previous studies in Chesapeake Bay tributaries suggested that denitrification was indeed an important process in their nitrogen budgets (Boynton et al. 1980; Jenkins and Kemp 1984; Twilley and Kemp 1986). Furthermore, a recent review of denitrification processes (Seitzinger 1988) indicated that some 50% of the nitrogen inputs from surrounding watersheds was removed through denitrification as nitrogen gas ($N_2$). Nitrogen gas is virtually unavailable to phytoplankton for uptake in estuaries (Day et al. 1988).

Numerous studies of estuaries worldwide have demonstrated the importance of recycled inorganic nitrogen compounds from sediments to overlying water as a source of nitrogen to support phytoplankton growth (Billen 1978; Blackburn and Henriksen 1983; Klump and Martens 1983; Nixon and Pilson 1983; Kemp and Boynton 1984; Christensen et al. 1987). For this reason, research activities have focused on the relative importance of sediment nitrogen cycling in the main stem of Chesapeake Bay.

Seasonal cycles (March through November) of nitrogen fluxes and transformation processes have been described for 10 m and 20 m (depth) stations in the Bay's mesohaline region (Figure 2, near station M4). Figure 11 summarizes rates for the 10 m station: ammonium and nitrate fluxes across the sediment-water interface were similar to those shown in Figure 6 for the mid-Bay station (20 m depth), where effluxes of ammonium from sediments follow the seasonal temperature cycle, peaking in August; nitrate fluxes are small except for the substantial uptake by sediments in spring. In contrast, nitrification and denitrification rates were highest in spring and fall, and they were virtually zero in summer.[5]

Denitrification measurements probably underestimated actual rates by about 30-60% because deficiencies in the methodology under high sulfide, low nitrate conditions at these sites (Christensen et al. 1989). The two processes were nonetheless highly correlated over the seasons, suggesting that denitrification was probably limited by nitrate produced in nitrification. Seasonal patterns of nitrification were, in turn, controlled by depth of oxygen penetration into the sediments (Kemp et al. 1990). Evidently, sediments at these sites were characterized by such high oxygen demand in summer that the oxygenated nitrification zone in surficial sediments was negligible (Henriksen and Kemp 1988); this was so even at the shallow station when relatively high oxygen concentrations were present in the overlying water.

Simple budgets of nitrogen fluxes and transformations were developed for this 10 m station in the mesohaline Bay in April, August and November (Figure 12). In April and November, inputs of particulate organic nitrogen to the sediment surface far exceeded outputs of ammonium and nitrogen gas (through denitrification) to overlying water; this imbalance was further exaggerated by the uptake of nitrate in April. In August, inputs to and outputs from the sediments were reasonably well balanced. For the 20 m station, similar nitrogen budgets indicate that efflux of solutes (primarly ammonium) generally exceeded inputs, especially in summer. The excess organic nitrogen inputs at the shallow site, compared to the deficiency estimated for the deeper site, represents further indirect evidence for bedload transport of particulate organic matter from shallow to deep areas of the Bay. These budgets also emphasize the relatively reduced role of nitrification and denitrification (calculated by budgeting the nitrate pools) compared to ammonium regeneration to overlying water, where the former constitutes only 10-20% of the latter (Kemp et al. 1990).

---

[5] These nitrogen transformation processes were measured by inhibitor techniques (N Serve [Henriksen et al. 1981] and acetylene [Sørensen 1978], respectively).

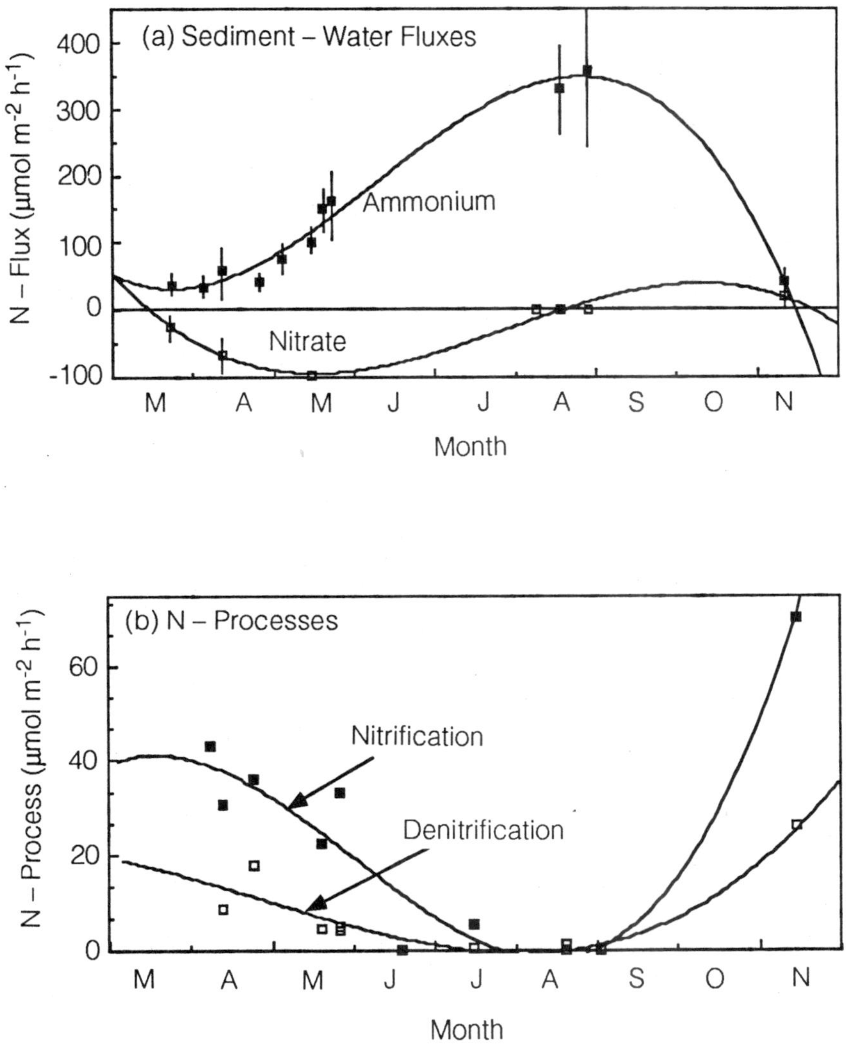


*Figure 11. Annual cycles (March to November) of (a) net sediment-water exchanges of ammonium and nitrate plus nitrite and (b) sediment nitrification and denitrification rates for a shallow (10 m) station in the mesohaline region of Chesapeake Bay (Kemp et al. 1990).*

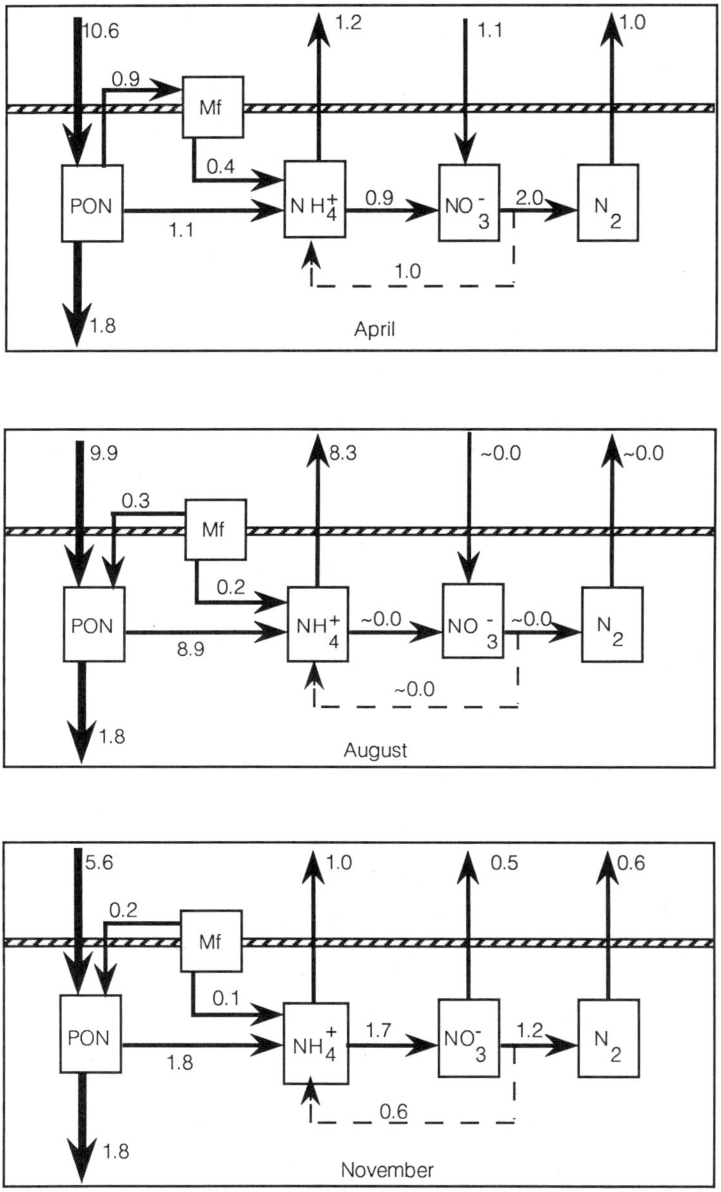


*Figure 12. Budgets of nitrogen transformation processes and net exchanges across the sediment-water interface for three periods at a shallow (10 m) station in the mesohaline region of Chesapeake Bay (Kemp et al 1989). Rates are in units of mmoles N $m^{-2}d^{-1}$. Nitrogen pools considered include: particulate organic N (PON); ammonium, nitrate, dinitrogen gas and macrofauna biomass (Mf).*

Recent measurements of nitrification and denitrification, as well as nitrate and ammonium fluxes, at stations all along the main salinity gradient of the Bay (Kemp unpublished), indicate somewhat different seasonal patterns for upper- and lower-Bay stations. Rates in the upper Bay were relatively high throughout the year, while nitrification (and associated denitrification) rates at the more saline stations appeared to be limited by low ammonium concentrations in late spring and summer. Preliminary time-space integrated estimates of denitrification for the whole main stem Bay suggest that nitrogen losses (i.e., nitrogen gas through denitrification) account for only 25 to 30% of the nitrogen inputs from the watershed. This is in contrast to the general pattern reported by Seitzinger (1988), where denitrification appeared to account for 50% (or more) loss of all nitrogen inputs to various estuaries. It appears that this relatively reduced importance of denitrification in the main Bay results from the decreased availability of dissolved oxygen in summer to support nitrification. When hypoxic conditions become chronic, normal nitrogen processing in sediments is suppressed and ammonium release predominates. This has the effect of reinforcing the hypoxic conditions by further supporting the cycle of algal production in overlying waters, followed by particulate organic matter deposition and consumption as well as oxygen depletion from bottom waters.

We can summarize pelagic-benthic nitrogen cycling at the 20 m mesohaline station site by comparing rates of nitrogen incorporation in phytoplankton growth to rates of organic nitrogen deposition to the sediments to rates of ammonium return from sediment processes back to the overlying water (Table 1). Mean data are provided for the period April to November from 1985 to 1987. Peak rates of organic nitrogen production occurred in mid summer, while deposition averaged over 1 to 2 month periods was relatively constant. As reported for other coastal sites, seasonality of ammonium regeneration from sediments corresponded closely to that for production (Kelly and Nixon 1984; Wassmann 1984). Deposition ranged from about 30 to 80% of production, and regeneration accounted for 40 to 210% of deposition. The overall averages of the four monthly mean rates suggest that 40% of the organic nitro-

Table 1. Summary of major benthic-pelagic exchanges of nitrogen over an annual cycle in the mesohaline region of Chesapeake Bay: production and depositon of particulate organic nitrogen (PON) and benthic regeneration of ammonium.

| Months | Plankton PON Production[a] | Net PON Deposition[b] | | Benthic $NH_4^+$ Regeneration[c] | |
|---|---|---|---|---|---|
| | (mg N $m^{-2}d^{-1}$) | (mg N $m^{-2}d^{-1}$) | %Prod | (mg N $m^{-2}d^{-1}$) | % Dep |
| Apr-May | 189 | 89 | 47 | 38 | 43 |
| Jun | 238 | 81 | 34 | 118 | 146 |
| Aug | 334 | 86 | 26 | 182 | 212 |
| Oct-Nov | 143 | 110 | 77 | 73 | 66 |
| Totals | 904 | 366 | 40 | 411 | 112 |

[a] Data adapted from Sellner et al. 1988.
[b] Data adapted from Boynton et al. 1988.
[c] Does not include nitrogen loss via denitrification.

gen production is deposited. Regeneration averaged more than 110% of the deposition, possibly indicating additional sources of organic nitrogen inputs to sediments such as lateral transport of resuspended material along the bottom.

## Interactions Between Phytoplankton and Benthic Macrofauna

Benthic invertebrate macrofauna are important components in coastal marine ecosystems such as Chesapeake Bay. These organisms constitute key food sources for many commercial finfish, including striped bass, spot, flounder, white perch and as valuable fisheries in themselves, in particular, blue crabs, oysters and soft clams. The soft bottom habitats of the Bay are dominated by polychaete worms and various bivalves (clams, for example) with crustacean amphipods also quantitatively important in some regions at different times of the year. Although other suspension-feeding macrofauna which attach to hard substrates — for instance, oysters, hydroids and barnacles — may have been dominant forms of benthic biomass in previous times, they now appear to be of secondary importance because of the generally impoverished oyster beds (Gerritsen et al. 1988).

Seasonal trends of macrofauna biomass and numerical abundances have been described for the upper and middle Bay's muddy and sandy sediment habitats (Holland 1988). While numerical abundances are similar for all areas, peak biomass values are an order of magnitude higher in the upper Bay because of the dense populations of molluscs (Holland et al. 1977). Similar patterns have been described for several major tributaries of the Bay which have also been studied extensively (Holland 1988). The importance of polychaetes increases down the Bay especially in muddy sediments. In fact, macrofaunal communities are dominated by large polychaetes in the polyhaline regions of the estuary (R. Diaz and L. Schaffner personal communication).

Seasonal maxima in macrofaunal abundance occur in June for most areas of the upper and middle Bay, followed by a precipitous decline in abundance and biomass with the onset of low oxygen conditions in summer, especially at the deeper muddy habitats. Various explanations have been postulated for the July decline. Anoxia, *per se*, is probably responsible at the deep (greater than 9 m) stations, while intensified predation in the shallower regions above the pycnocline (Holland et al. 1980) may account for the summer macrofaunal declines there (Kemp and Boynton 1981). With the crash of the spring diatom bloom in May (Malone, this volume), the marked decline in particulate organic matter deposition (Figure 3) may also result in food limitations for the benthos in some areas of the Bay (Marsh and Tenore 1990). The higher carbon to nitrogen (C:N) and carbon to chlorophyll *a* (C:Chl) ratios of particulate organic matter deposited in summer indicate a lower nutritional content of food available to support benthic macrofauna after June (Christensen and Kanneworff 1985; Gardner et al. 1989).

Feeding guilds of benthic macrofauna grade sharply from predominantly "head up" feeders, for example, the clam *Macoma balthica* and the polychaetes *Nereis succinea* in the upper and middle Bay to predominantly "head down" feeders, for example, the polychaete *Macroclymene zonalis* in the lower Bay; predatory forms are more important in sandy sediments of the middle and lower Bay (Holland 1988). This transition in predominance of head-up to head-down macrofaunal feeding types roughly parallels the decreasing organic content and increasing redox conditions of sediments from the upper-middle to lower Bay. Presumably, low redox conditions and relatively high rates of particulate organic matter deposition in the northern regions of the Bay favor infaunal species which respire in the overlying water and feed at the sediment-water interface. In contrast, the lower inputs of particulate organic matter to sediments of the lower Bay select for animals capable of reworking the organic pools in the relatively oxidized sediments (Jumars and Fauchald 1977).

To the extent that food availability limits growth of benthic animals, expectations are that on broad time and space scales biomass and secondary production of macrofauna would be directly proportional to deposition rates of particulate organic matter from overlying waters to sediments (Rowe 1971; 1985; Parsons et al. 1979; Josefson et al. 1987; Grebmeier et al. 1988). In Chesapeake Bay, contemporaneous seasonal data for particulate deposition (i.e., sediment traps) and benthic biomass are available for only one mesohaline site (Figure 2, Sta M4); however, on seasonal time scales no significant correlation between particulate organic matter deposition and benthic biomass could be found for this station. The absence of a relationship here is largely a consequence of the seasonal loss of benthic macrofauna associated with oxygen depletion from bottom waters (Holland et al. 1977; Kemp and Boynton 1981). Therefore, it would be more likely to find a relationship between plankton deposition and benthic biomass by comparing seasonal or annual mean values among different stations with similar seasonal patterns of dissolved oxygen.

Since there is only one sediment trap site in Chesapeake Bay, it is impossible to make spatial comparisons of relations between macrofaunal abundance and particulate organic matter deposition. Contemporaneous measurements are available, however, for phytoplankton production and abundance as chlorophyll *a* (Sellner personal communication) and biomass of benthic macrofauna (Holland personal communication) at several stations in the mainstem and tributaries of the Bay. Statistically significant correlations were observed between annual mean biomass of polychaetes (g ash free dry weight $m^{-2}$) and euphotic zone chlorophyll *a* (mg $m^{-2}$) in spring among these Bay sites when the benthic data were partitioned to separate stations experiencing summer hypoxia or anoxia in bottom waters (Figure 13). The slope of the relation for the hypoxic stations is, however, an order of magnitude less than that for the oxic stations, indicating the stress effects of seasonal oxygen depletion on the metazoan benthic community (Kemp and Boynton 1981). No significant correlation was observed between values for benthic macrofauna and phytoplankton abundance in summer. The absence of such relations emphasizes the importance of the spring deposition events in supporting these benthic populations (Parsons et al. 1979; Townsend and Cammen 1988).

In the above analysis, the other main group of organisms composing the majority of benthic biomass in Chesapeake Bay — bivalves — was purposely omitted from the macrofauna because these animals, which are primarily suspension feeders, are not just passive recipients of particulate organic matter deposition. In fact, their active grazing can directly reduce plankton abundance and production (Officer et al. 1982; Cloern 1982). When the biomass of suspension-feeding bivalves is compared with plankton abundance and production, the patterns are quite different than those for polychaetes (Figure 14). Note that the independent variable (x-axis) has been reversed between Figures 13 and 14 to emphasize a shift in the dominant direction of causality. An inverse relationship between spring biomass of bivalves and annual mean values of euphotic-zone chlorophyll *a* and phytoplankton production was obtained for the monitoring stations which do not experience seasonal anoxia. This pattern suggests that phytoplankton are unaffected by bivalve abun-

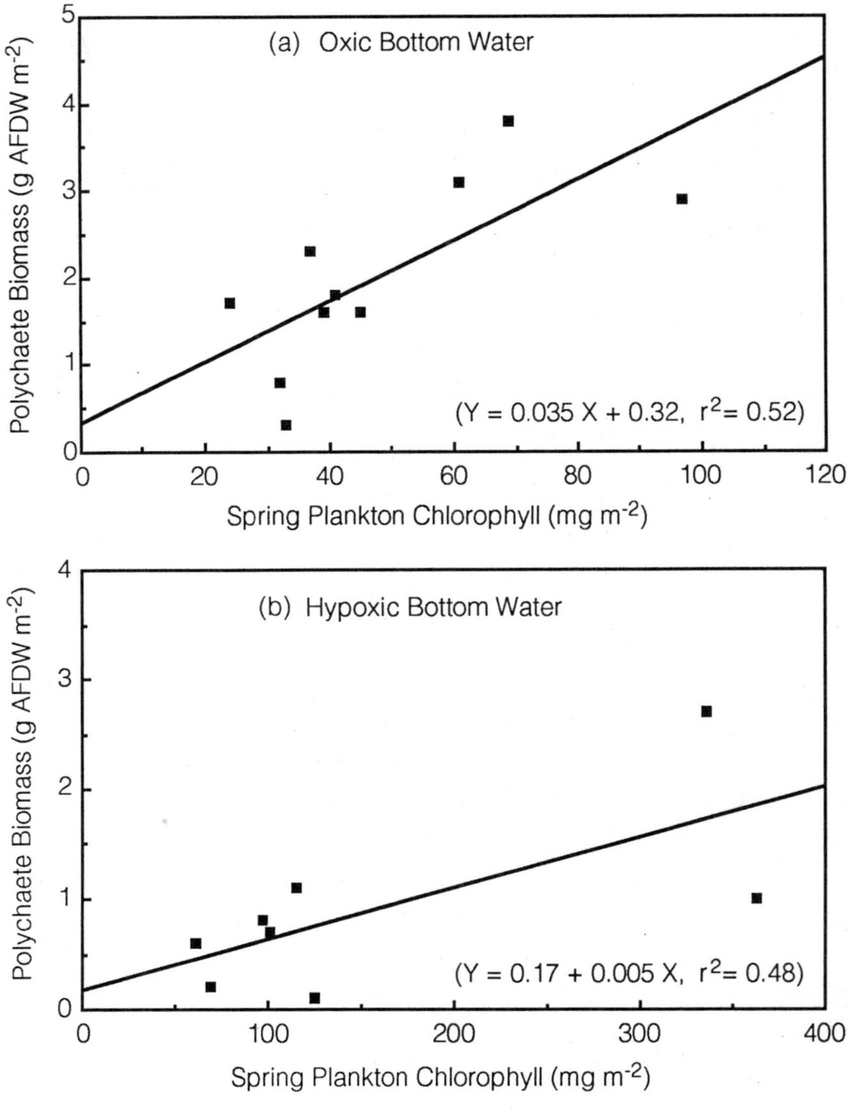


*Figure 13. Correlations between annual mean macrofaunal biomass (excluding clams) versus (a) spring peak phytoplankton production and (b) spring peak phytoplankton chlorophyll* a. *Each point represents a different station in the main stem or tributaries of the Bay (data taken from Sellner 1988 and Holland 1988).*

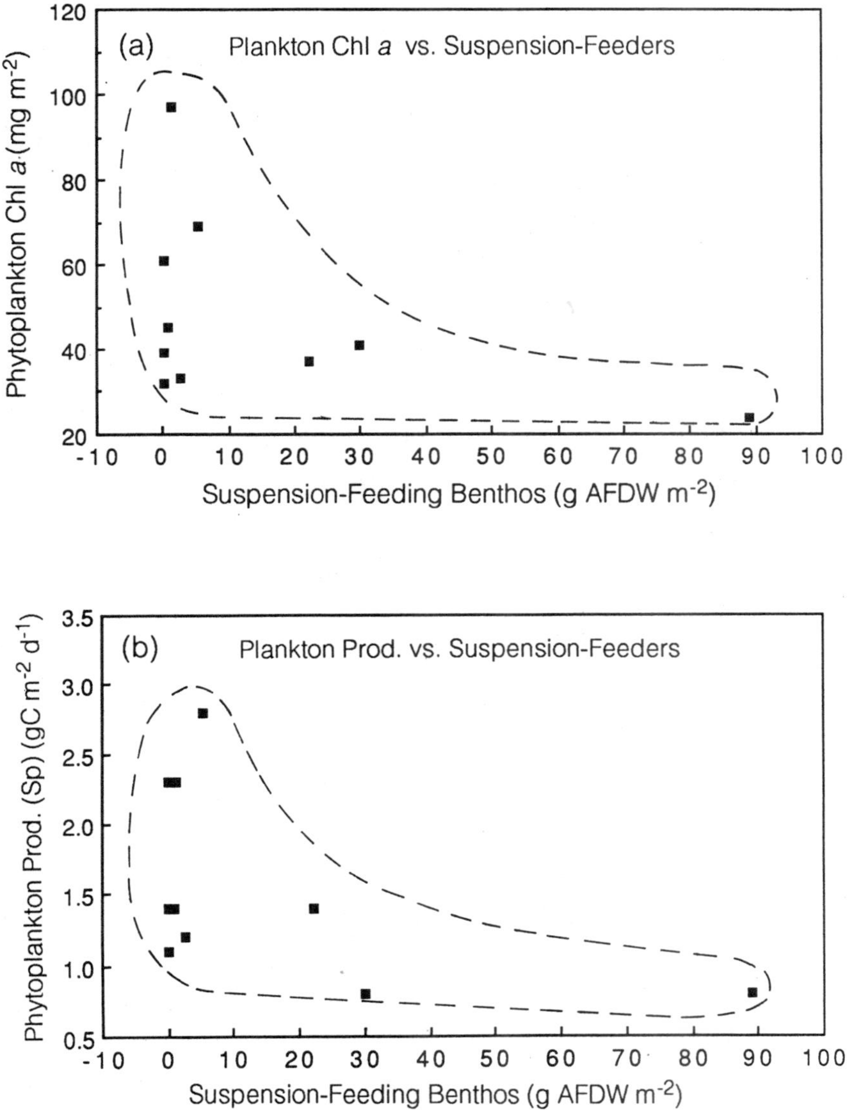


*Figure 14. Relations between annual mean biomass of suspension-feeding macrofauna versus (a) spring peak phytoplankton chlorophyll* a *and (b) spring peak phytoplnkton production. Each point represents a different station in the main stem or tributaries of the Bay (data taken from Sellner 1988 and Holland 1988).*

dance for most stations in the middle to lower regions of the Bay and its tributaries, whereas phytoplankton are markedly reduced at the upper estuarine stations (mainstem Bay, Patuxent and Potomac rivers) which support abundant clam populations. Light limitation on algal growth may also contribute to low phytoplankton abundance in low salinity Bay regions.

Previous analyses of the impact of the asiatic clam, *Corbicula fluminea*, on phytoplankton in the upper Potomac River estuary by Cohen et al. (1984) have reported similar inverse relations between plankton and clam abundances. These investigators have also demonstrated experimentally the ability of the animals to dramatically graze plankton from the water column. While the invasion of this exotic species into the upper Potomac may represent an unusual event, recent calculations (Gerritsen 1988; Gerritsen et al. 1988) suggest that clam populations in the fresh water regions of other tributaries as well as the mainstem Bay are sufficient to significantly reduce phytoplankton abundance.

Prior to the latter part of the nineteenth century, the largest population of suspension-feeding animals in Chesapeake Bay was probably the American oyster, *Crassostrea virginica.* Intensive harvesting of oysters since the turn of the century and more recently, disease has resulted in dramatic (100-fold) declines in their abundance throughout the Bay (Kennedy and Breisch 1981). Newell (1988) has, therefore, postulated that the significant decrease in grazing by oyster populations across the Bay is responsible for much of the increase in phytoplankton biomass since the 1930s which has otherwise been attributed to eutrophication of this estuary (Boynton et al. 1982; Officer et al. 1984). Because clam abundances are so large in the upper Bay regions, it is difficult to ascertain whether the total activity of suspension-feeding benthic macrofauna has decreased with the decline of oysters (Gerritson et al. 1988). It is clear, however, that spatial distributions of suspension-feeders have changed significantly. An important consideration is the effect of reef-forming animals like oysters on the increasing transport of plankton to the bottom by reducing the height of the benthic boundary layer (Frechette et al. 1989). On the other hand, increased recycling of nutrients and consumption of oxygen associated with the metabolism of suspension feeding benthic animals themselves and with their fecal and pseudofecal matter (see Jordan 1985) could serve to stimulate phytoplankton growth and oxygen depletion during summer (Doering et al. 1987). Thus, it remains an open question as to the ultimate effect of shifts in abundance of suspension-feeding benthic macrofauna on the "greening" of the Bay with increased phytoplankton and/or the depletion of oxygen from its bottom waters.

## *Comparisons of Macrofaunal Interactions with Other Systems*

Parsons et al. (1979) have reported a clear relationship between spring biomass of phytoplankton and peak macrofaunal abundance across a wide range of coastal marine environments (Figure 15). Presumably, this kind of correlation would be even stronger if rates (i.e., particulate organic matter deposition and macrofaunal

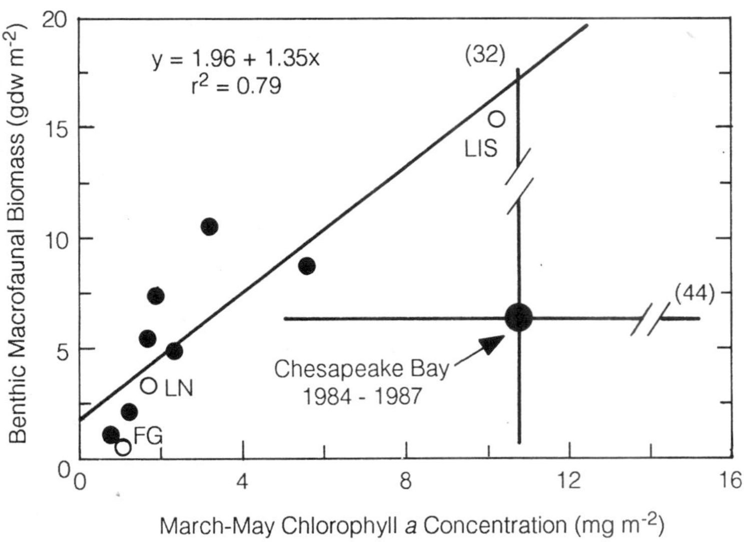


*Figure 15. Summary of benthic macrofaunal biomass versus spring chlorophyll a for various marine sites reported in the literature (Parsons et al. 1979), with values for annual means and ranges from Chesapeake Bay (Holland 1988; Sellner et al. 1988) presented for comparisons.*

production) were compared rather than stocks, though such data are rarely available for annual cycles at a given site. No significant relation from these data was obtained when summer phytoplankton biomass were considered instead of spring values (B. Hargrave personal communication). Other investigators have emphasized the importance of the spring diatom bloom as the primary source of food to sustain growth and recruitment of benthic macrofauna in temperate marine regions (Graf et al. 1982; 1983; Hargrave and Phillips 1986). The importance of the spring bloom in providing large quantities of high quality food for the benthos lies in the fact that much of this spring bloom is deposited as intact, ungrazed diatoms (Smetacek 1984). In fact, Townsend and Cammen (1988) have postulated that the relative magnitude of benthic invertebrate production and associated success of juvenile demersal fish recruitment for a given year in the Gulf of Maine depends on the timing of the spring bloom, where maximum deposition and recruitment success occur for early blooms which precede the period of intense zooplankton grazing. Indeed, sinking particulate organic matter has maximum nutritional value for growth of benthic animals when dominated by intact diatoms (Christensen and Kanneworf 1985; Gardner et al. 1989).

In nutrient-rich stratified coastal systems such as Chesapeake Bay, the seasonal depletion of oxygen from bottom waters in late spring to early summer complicates this relation between phytoplankton and benthic macrofauna (e.g., Figure 13). The mean and range of spring values for phytoplankton chlorophyll *a* and macrofauna biomass at all Chesapeake Bay stations (see Figure 2) are also plotted on Figure 15. Given the relatively high chlorophyll *a* levels in the Bay, mean macrofauna biomass is much less than expected from the relation. This deficiency in the abundance of Chesapeake Bay macrofauna, compared with the high phytoplankton production, again emphasizes the negative effects of eutrophication and associated oxygen depletion from the Bay's bottom waters.

Nixon (1988) recently pointed out the relatively low rate of fisheries yield per unit primary production for fresh water ecosystems compared to coastal marine regions. Previous studies have established strong correlative relations between nutrient loading and phytoplankton production (Schindler 1981) and between benthic macrofaunal abundance and fisheries yields for various lakes. Nixon (1988) attributed this pattern of higher relative fish yields per unit primary production in marine ecosystems to the seasonal anoxia in bottom waters of most temperate lakes compared to the oxic bottom waters of most tidally mixed estuaries. However, in estuaries such as Chesapeake Bay which undergo seasonal anoxia, this reasoning implies that demersal fisheries may also be significantly depressed, due to the impact of anoxia on benthic-pelagic coupling.

## Role of Benthic and Pelagic Processes in Oxygen Depletion

During spring in the mesohaline region of the Chesapeake Bay main stem, vertical stratification of the water column combines with springtime warming and the deposition and consumption of the diatom bloom to produce a rapid decline in oxygen concentration in the bottom waters (Taft et al. 1980; Malone, this volume). Similar patterns of seasonal anoxia have been reported for numerous coastal marine systems (Richards 1965; May 1973; Jørgensen 1980; Falkowski et al. 1980; Caumette et al. 1983). Vertical density stratification, which reduces the exchange of oxygen between the upper and lower layers of the water column, plays a key role in this phenomenon (Turner et al. 1987). Therefore, meteorological and hydrological processes strongly influence the magnitude of anoxic events in Chesapeake Bay (Seliger et al. 1985; Seliger and Boggs 1988; Boicourt, this volume). However, it is the production, deposition and consumption of organic matter which fuels this process each year (Malone et al. 1986). In principle, the elevated inputs of nutrients in eutrophic coastal ecosystems could increase the spatial and temporal dimensions of bottom water anoxia by increasing algal production (Boynton et al. 1982; Malone et al. 1988). Recent evidence suggests that this may be the case for Chesapeake Bay (Officer et al. 1984), and ecosystem simulation modeling studies corroborate the rationale for this relationship (Kemp and Randall 1988).

The mean and range of time delays between organic production in the upper layers and subsequent consumption (and associated oxygen depletion) in the lower water column and benthos are poorly understood. How much organic production is carried over between seasons and years to support subsequent consumption processes? This question is important from a management standpoint because of a need to know how rapidly oxygen concentrations and related ecological processes will respond to controls on eutrophication. Understanding how total oxygen consumption is divided between planktonic and benthic processes would provide some insights on this question because there tends to be a more rapid coupling between production and planktonic respiration than between production and benthic respiration (Oviatt et al. 1986). A portion of the benthic metabolism of coastal marine systems involves complex sequences of microbial and geochemical processes which may require considerable time to complete (Berner 1980). In addition, the relative contributions of biological processes (production, respiration) and physical processes (stratification, transport) to oxygen depletion is of concern to managers because most control measures for eutrophication (e.g., reduced nutrient inputs) involve only the biological factors.

Both planktonic and benthic respiration processes in the Bay increase rapidly with vernal warming in spring. The rate of increase between March and June is more rapid for planktonic respiration, peaking in early July compared to an early August maximum for benthic oxygen consumption (Kemp unpublished data). Pre-incubation filtration experiments indicated that most of the plankton respiration in bottom waters was associated with bacteria and protozoa (< 3 μm) (Jonas, this volume). Similar microbial dominance of planktonic respiration has been reported for other coastal systems (Williams 1981; Hopkinson et al. 1989). The high August rates associated with the benthos actually occur in the water column at the pycnocline (oxycline) via oxidation of sulfides produced in sulfate reduction and diffusing from sediments to overlying water (e.g., Indrebo et al. 1979; Dyrssen 1986). These rates of sulfide efflux from sediments do not occur until several days after oxygen has been depleted from bottom waters. Considering the mean depth of the bottom layer in this region of the Bay, the relative magnitudes of benthic and planktonic rates can be compared over the spring summer period (Figure 16a). Although benthic rates exceeded planktonic rates in early April and late August in 1986, for most of this period planktonic respiration was 1.5 to 2 times greater.

The magnitude of physical transport processes in the seasonal budget of oxygen is more difficult to ascertain (Boicourt, this volume). However, rates can be inferred from careful analysis of budgets of oxygen inputs to and outputs from the upper and lower layers of an average square meter of water column in the mesohaline region (e.g., Kemp and Boynton 1980). For the present analysis of 1986 data, it was assumed that net longitudinal exchanges of oxygen were negligible in the upper layer (Kemp, unpublished data). This assumption is supported by the absence of any consistent longitudinal gradient of oxygen content (as % saturation) of the surface waters. Vertical exchange between layers was then calculated as the difference between observed rates of change in oxygen and those calculated as the sum of all measured rates. Horizontal oxygen transport was, however, allowed in the budget for the bottom layer (Kuo and Neilson 1987), and net rates were calculated as above, by difference. Table 2 summarizes details of these calculations for April, May and August. By way of an independent check, the fact that calculated rates of vertical and horizontal oxygen transport were significantly correlated ($r^2 > 0.60$) to their respective oxygen concentration gradients lends credibility to this indirect approach.

For purposes of comparing biological and physical processes, monthly means for the algebraic sum of oxygen fluxes associated with physical transport were plotted against mean total respiration rates for the bottom layer from March to October in 1986. It can be seen that respiration exceeds oxygen replenishment by physical transport for most of the spring and early summer, but especially in

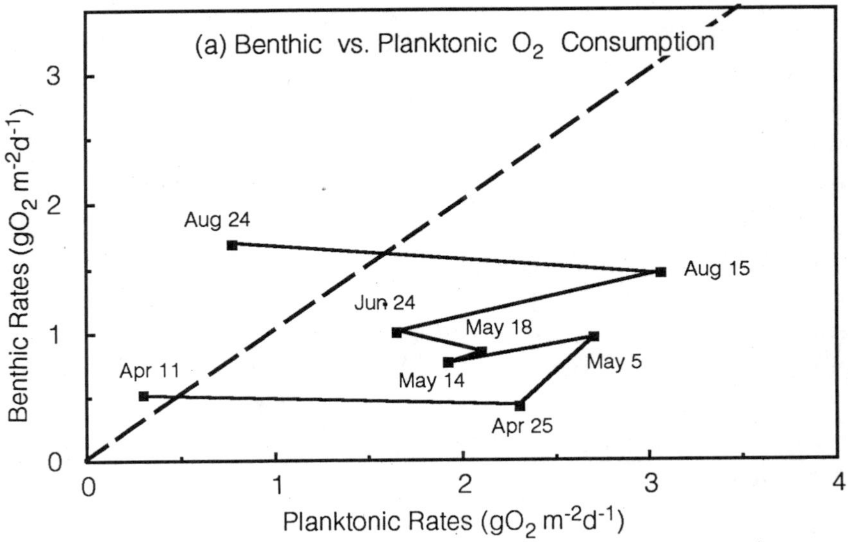


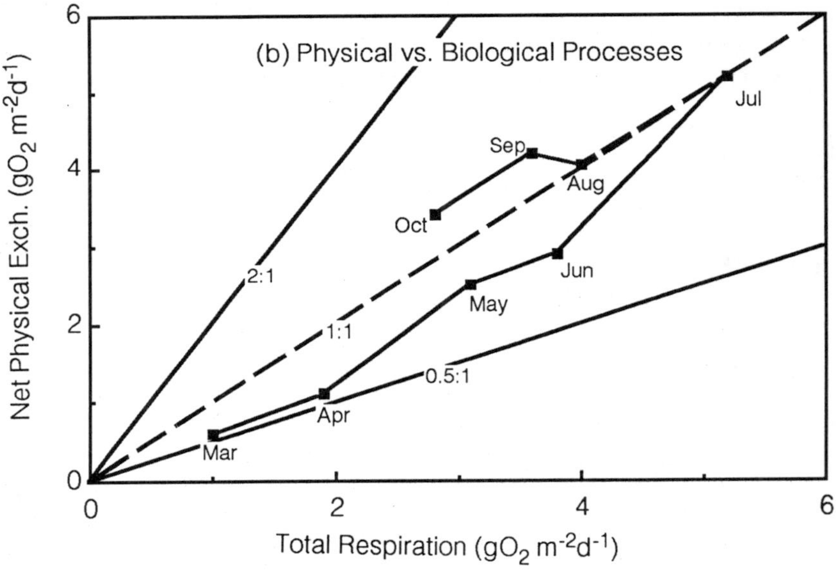


*Figure 16. Seasonal sequence of (a) benthic versus planktonic oxygen consumption rates (April-August 1986) and (b) oxygen consumption by biological processes versus net oxygen exchange via physical processes (March-October 1986). Taken from Kemp et al. (unpublished).*

Table 2. Summary balance of oxygen fluxes ($gO_2\ m^{-2}d^{-1}$) for upper and lower layers of water column in the mesohaline reach of Chesapeake Bay during Spring and Summer.

| Water Column Layer<br>Oxygen Fluxes | April | May | August |
|---|---|---|---|
| Upper Layer: | | | |
| Production[a] | +5.4 | +6.8 | +9.4 |
| Plankton Respiration[b] | -3.0 | -4.3 | -5.3 |
| Air-Water Exchange[c] | -1.2 | -1.2 | -1.2 |
| Temporal Change | 0 | -0.1 | -0.3 |
| X-Pycnocline Exchange[d] | -1.2 | -1.4 | -3.2 |
| Lower Layer: | | | |
| X-Pycnocline Exchange[d] | +1.2 | +1.2 | +3.2 |
| Plankton Respiration | -1.3 | -2.2 | -1.9 |
| Benthic Respiration[e] | -0.5 | -0.9 | -1.6 |
| Temporal Change | -0.8 | -0.6 | 0 |
| Net Horizontal Exchange[f] | -0.1 | +1.1 | +0.8 |

[a] Gross $O_2$ production estimated from net daytime (apparent) production plus half of the dark respiration (assuming day and night respiration rates are equal and 12 h daylight).

[b] Dark respiration of whole water; assumes upper layer height of 8 m and mean depth of lower layer equal to 6 m, except in August when lower layer is anoxic and pycnocline thickness of 3 m is used.

[c] Assumes air-water exchange coefficient equal to 1.0 $gO_2\ m^{-2}h^{-1}$ at 100% saturation deficit (Kemp and Boynton 1980); diel mean % $O_2$ saturation taken at 105%.

[d] Net $O_2$ exchange across pycnocline estimated as residual term in upper layer $O_2$ balance.

[e] Benthic $O_2$ consumption estimated as mean of rates measured at 10 m and 20 m depth; rates at 20 m include sulfide efflux in August.

[f] Net longitudinal $O_2$ exchange estimated as residual term in lower layer $O_2$ balance.

April (Figure 16b). During most of the year, the balance between physical and biological processes does not stray far from the 1:1 line, which is intersected in July and continually exceeded until October.

These data illustrate the fact that biological and physical processes of oxygen transformation and transport are poised in near balance throughout the period from spring to autumn. This is not surprising because of the interdependence of the two kinds of processes. Physical transport depends on concentration gradients which are created only by biological processes. On the other hand, at low concentrations of oxygen, biological processes follow first-order kinetics and are directly proportional to the rate of physical transport of oxygen. This balance is analogous to that in the diffusion-reaction model which characterizes many biogeochemical processes.

In summary, we find that planktonic respiration is quantitatively more important than benthic processes for most of the spring period of declining bottom water oxygen. Benthic respiration is, however, more significant at the beginning and end of the anoxia period. Biological consumption of oxygen in the bottom layer is reasonably balanced with physical replenishment processes throughout the period between March and October, such that small changes in either could have marked effects on the magnitude and timing of seasonal anoxia. Numerical ecosystem modeling studies of this system have illustrated the importance of this balance between biological and physical processes: in the absence of vertical wind mixing in summer, anoxia persisted until November, while removal of external nutrient inputs results in virtual elimination of anoxia within one year (Kemp and Randall 1988).

## Effects of Eutrophication on Benthic-Pelagic Coupling

Limnologists have long recognized that for many lakes, the process of continual nutrient enrichment is part of a natural sequence in the long-term history of these aquatic ecosystems (Hutchinson 1974). It is well documented that the anthropogenic acceleration of this process — that is, nutrient enrichment in river flow and airborne deposition — tends to produce a range of dramatic changes in the structure and function of lake ecosystems, including depletion of bottom water oxygen, losses of submersed vascular plants and alterations in food chains and associated fisheries (e.g., Edmondson 1969; Lehman 1988). In contrast, for coastal marine ecosystems evidence has been scant that eutrophication is a natural biogeochemical process, though there is growing recognition that the widespread pattern of rapidly increasing anthropogenic nutrient inputs to coastal marine waters is resulting in pronounced alterations of organism abundances and ecological interactions (Nixon et al. 1986).

Indirect evidence exists that in Chesapeake Bay responses of ecological processes to changes in nutrient loading from the surrounding watershed may be relatively more pronounced than for many other estuaries. In fact, significant positive relationships between rates of phytoplankton production and nitrogen inputs (Boynton et al. 1982) were observed both for Chesapeake Bay and its tributaries and for a variety of other estuaries combined together (Figure 17a). We have postulated that the higher production per unit nutrient inputs in the Bay may indicate greater retention and recycling of these inputs (Kemp and Boynton 1984). Since most of the other estuaries in Figure 17a are vertically well mixed, it can be inferred that the efficient trapping of particles in the Bay's two-layer gravitational circulation may be an important factor (Boicourt, this volume). Evidently, this nutrient retention occurs at interannual time scales (Boynton et al. 1982). For instance, a significant correlation between river flow, the primary source of nutrients to the Bay, and phytoplankton was observed only when a two-year running mean flow was used, implying a residual effect of one year's nutrient inputs on the next year's production (Figure 17b) (Sellner personal communication). Because this same physical mechanism of vertical stratification which evidently leads to enhanced nutrient retention, also contributes to reduced reoxidation of bottom waters, estuaries such as the Chesapeake may be more susceptible to increased anoxia in response to eutrophication.

While we have already described several observed or postulated changes in Chesapeake Bay's benthic-pelagic interactions which

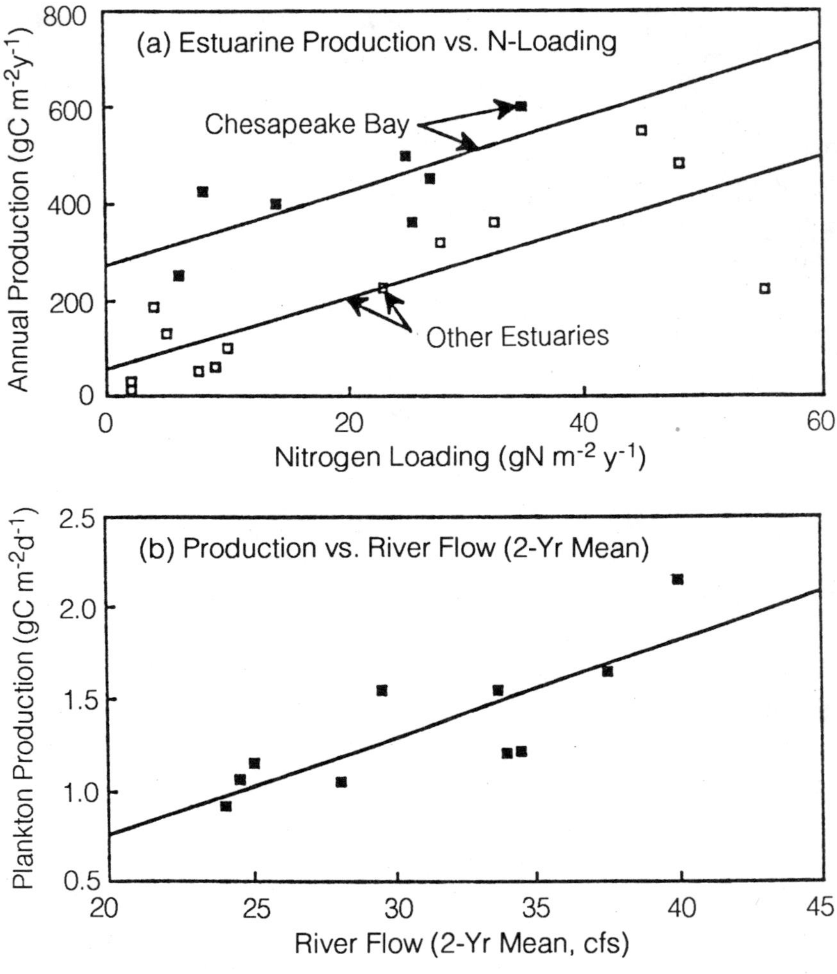


*Figure 17. Phytoplankton production versus nutrient loading from river flow: (a) annual production versus nitrogen loading rates for Chesapeake Bay and other estuaries (Boynton et al. 1982); (b) annual mean daily production versus two-year running mean of river flow for the mesohaline region of Chesapeake Bay (Sellner unpublished data).*

might be attributable to eutrophication, here we summarize these relationships between coastal eutrophication and benthic-pelagic couplings in an integrated framework. Our discussion focuses on the following major mechanisms by which nutrient enrichment might alter key processes of benthic-pelagic coupling: (1) effects on plankton trophic dynamics and resulting particulate organic matter deposition; (2) effects on benthic macrofaunal abundance and demersal fisheries; (3) effects on rates and temporal scales of benthic regeneration processes. The causal connections and pathways by which these mechanisms occur can be illustrated by referring back to the conceptual scheme presented in Figure 1.

### *Planktonic Food-Chains and Particulate Organic Matter Deposition*

Particulate organic matter can sink from the upper water column toward the sediment surface in a number of forms: as intact algal cells; as zooplankton fecal pellets; as carcasses of dead animals; and as miscellaneous plant debris. The deposition of particulate organic matter from overlying water to sediments can occur via passive sinking or through the active pumping by benthic suspension feeders. As suggested in Figure 1, the two primary pathways of passive sinking, as algal cells or zooplankton feces, are regulated by the magnitude of primary production and by trophic interactions in the plankton community.

Hargrave (1973) reported benthic community metabolism was significantly correlated with the ratio of primary production to mixed layer depth; this correlation was characterized by a power function (exponent = 0.39) for a variety of coastal sites. The fact that the exponent of this relation is less than unity implies that as primary production over the mixed layer increases (as might be the case with eutrophication), the relative importance of benthic metabolism in consuming that production decreases. Here, it is assumed that benthic metabolism is directly proportional to particulate organic matter deposition, with no change in the fraction buried. This same pattern of decreasing proportion of total oxygen consumption by benthic — compared with planktonic — processes was seen along an experimental nutrient enrichment gradient in the MERL mesocosms (Oviatt et al. 1986).[6] In addition, a similar relation was also evident over an annual cycle in the Georgia Bight, where benthic respiration was proportionately less important relative to planktonic respiration during periods of peak production (Hopkinson 1987).

Several investigators have suggested that the specific nature of plankton trophic interactions can greatly affect the quantity and quality of particulate organic matter deposition to the benthos in coastal ecosystems (Smetacek 1984). For example, Hargrave (1973) postulated that such a pattern (described in previous paragraph) might be attributable to the fact that zooplankton communities were more efficient in consuming the relatively dense assemblages of phytoplankton characterizing eutrophic waters. Alternatively, any increase in the importance of bacterial and protozoan respiratory consumption of organic matter would result in a relative reduction in particulate organic matter deposition (Williams 1981), while increased grazing by certain gelatinous zooplankton might lead to increases in deposition (Andersen and Nival 1988).

### *Benthic Macrofauna and Demersal Fisheries*

Although eutrophication of coastal waters may result in an increase in the relative proportion of phytoplankton production consumed by planktonic as opposed to benthic processes, the absolute rates of particulate organic matter deposition will increase with eutrophication. In systems with vertically well mixed water columns, this increase in organic food available to the benthos would support greater growth of macroinvertebrates and demersal fish (e.g., Figure 13a). Indeed, higher abundance of benthic macrofauna (especially polychaetes) was observed in response to fertilization in the well-mixed overlying water columns of the MERL mesocsosms (Grassle

---

[6] The Marine Ecosystem Research Laboratory (MERL) is a facility located at the University of Rhode Island designed around fourteen cylindrical mesocosms (5.5 m in depth and 1.8 m in diameter). These mesocosms are set up and maintained as experimental ecosystems, and are used to conduct studies on the behavior of natural systems and their response to perturbations.

1985). However, in estuarine regions with vertically stratified water columns, increased phytoplankton production associated with eutrophication can lead to oxygen depletion, resulting in a decrease of benthic animals. In the Baltic Sea, for instance, increased algal production and deposition between the 1930s and the 1970s, presumed to be associated with eutrophication, resulted in different changes in benthic macrofaunal abundance within different depth regimes. The shallow regions not subject to hypoxia experienced macrofaunal increases, while in the deeper parts of the Baltic macrofauna decreased during this time (Cederwall and Elmgren 1980). Likewise, because much of Chesapeake Bay is vertically stratified, it is reasonable to assume that benthic macrofaunal production is inhibited in some regions and that the continuing trends of increasing eutrophication will lead to further losses of production for benthic invertebrates and associated demersal fisheries.

### *Rates and Temporal Scales of Benthic Nutrient Recycling*

Each year nutrient inputs to temperate estuaries such as Chesapeake Bay directly support organic production in the spring and indirectly support, through benthic recycling processes, continued high production in the summer. It is the benthic recycling processes which provide nutrients for most of the estuary's production in the warmer seasons (Kemp and Boynton 1984). For nitrogen, high rates of benthic recycling lead to large, rapidly exchanging sediment pools of ammonium in summer. If oxygen is available, a substantial portion of this ammonium will be oxidized to nitrate, which is in turn reduced to nitrogen gas ($N_2$) by nitrifying and denitrifying bacteria, respectively. Thus, high rates of nitrification-denitrification will occur in summer but only *under conditions of oxygenated bottom waters.* For many coastal ecosystems with well-mixed water columns, approximately half of the nitrogen entering from the watershed is transformed as nitrogen gas via nitrification-denitrification processes (Seitzinger 1988). For stratified eutrophic estuaries such as portions of the upper Chesapeake Bay, which experience severe hypoxia in bottom waters, this coupled nitrification-denitrification process is inhibited because of the lack of oxygen in bottom waters (Kemp et al. 1990). As a consequence, less than 30% of the nitrogen inputs to the mainstem Bay are removed as nitrogen gas. Therefore, while increased rates of nitrogen loading to the Chesapeake Bay have contributed to expanding eutrophication effects, associated hypoxic conditions appear to be resulting in an inhibition of nitrification and denitrification, which has the positive feedback effect of further stimulating the eutrophication process by increasing the retention and recycling of ammonium to support more phytoplankton production in summer.

Recent evidence suggests that interannual variations in benthic metabolic processes are proportional to changes in nutrient inputs to the Bay. Long-term records for these important ecological processes are not available to compare, for example, rates prior to European settlement with twentieth century rates. However, because river flow, which is the primary source of nutrients to the Bay, varies markedly from year to year, correlations between river flow and ecological processes during modern times can be used to infer relations over longer time-scales. For example, for the six years of data for which summer rates of benthic phosphorus recycling were available in the lower Patuxent River estuary, there was a significant relation to river flow in the preceding spring (Figure 18). Similarly, a continual decrease in rates of sediment oxygen consumption observed throughout the Bay main stem between 1985 and 1988 (Figure 19a) might be interpreted as a response to changes in plankton production. In fact, river inputs of nutrients in 1984 were among the highest on record, while the following four years were characterized by severe drought conditions with relatively little nutrient delivery to the Bay. One interpretation of this pattern is that a residual nutrient pool from inputs in 1984 contributed to recycling of nutrients in 1985 (Figure 17) to support relatively high plankton production (Malone et al. 1988) and benthic community metabolism (Figure 19b, c) in 1985. Consistent with this interpretation is the fact that despite the extremely low riverflow in 1985, bottom water anoxia in the mesohaline Bay was more severe in that year than in subsequent years (1986-1988), again suggesting a residual effect from 1984.

This concept that processes of benthic-pelagic coupling can be affected in one year by nutrient inputs from previous years has significant implications for strategies to manage nutrient waste inputs to the Bay. It is important to understand the extent to which manifestations from past eutrophication will continue to be experienced in the future even after reductions in nutrient inputs. Direct experiments have shown that particulate organic matter inputs to sediments will continue to produce high rates of benthic metabolism and nutrient recycling for several months but are generally dissipated to pre-treatment background levels in less than a year (Kelly and Nixon 1984; Rudnick and Oviatt 1986). These direct relationships between nutrient inputs in a given year and benthic recycling processes within that year or the next (Figures 18 and 19) suggest that most of the effects of nutrient loading are not retained for more than a year or two. It is still unclear, however, what is the extent to which small amounts of organic matter — retained in

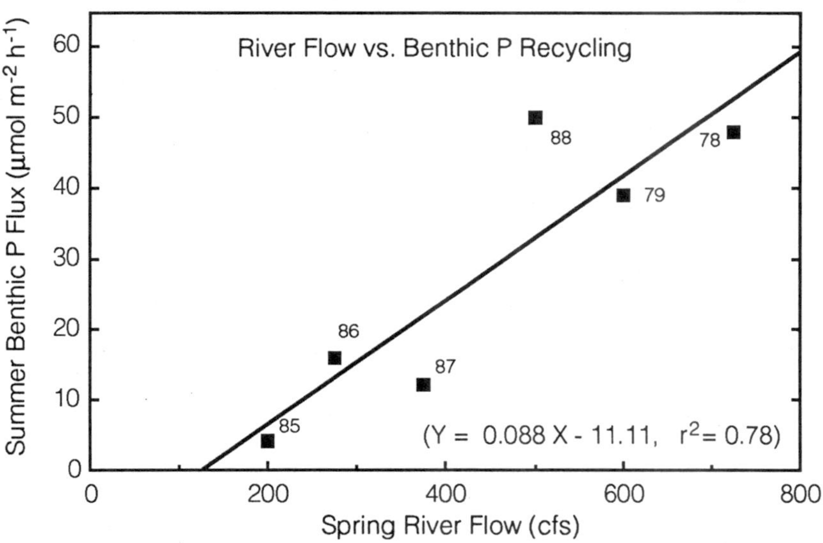


*Figure 18. Relationship between fluxes of dissolved inorganic phosphorus across the sediment-water interface at a station in the lower estuary and average spring (March-May) river flow of the Patuxent River; numbers next to points indicate years (Boynton et al. unpublished data).*

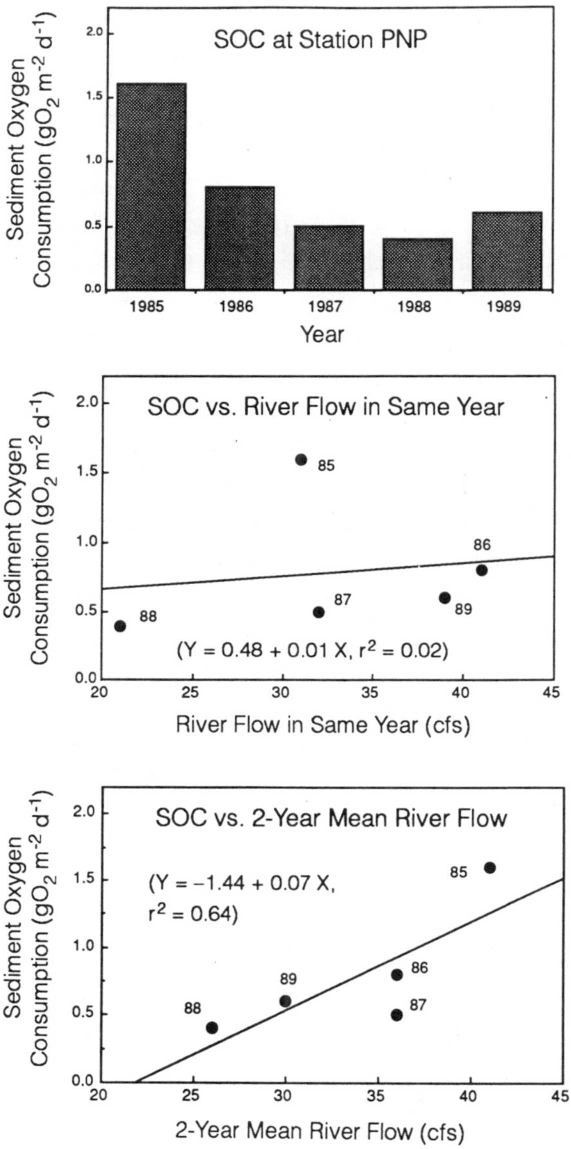


*Figure 19. (a) Annual mean rates of sediment oxygen consumption (SOC) at a station in the Chesapeake Bay main stem near the mouth of the Potomac River for 1985-1988; (b) correlation between SOC and river flow for each year; (c) correlation between SOC and two-year running mean of river flow in given year and that in previous year.*

sediments over decades and decomposing at slow rates — can contribute to the long term benthic process of oxygen consumption and nutrient recycling. With increasing eutrophication, the proportion of particulate organic matter deposition which is buried may well increase.

## Research Needs and Management Implications

During the last decade a great deal has been learned concerning benthic-pelagic coupling in coastal and estuarine systems including Chesapeake Bay. As is generally the case, the results now available pose additional specific questions that should be addressed to further our understanding and ability to predict aspects of estuarine ecosystem dynamics.

### *Research Needs*

At this time there is considerable uncertainty with regard to relations between plankton trophic dynamics and deposition rates of particulate organic matter. For example, our current understanding is that the spring bloom in the central regions of the Bay largely sinks as intact cells with little exposure to grazing in the water column. In contrast, summer phytoplankton stocks appear heavily grazed and large percentages of organic production are metabolized in the water column. The reasons for and implications of this seasonal shift are not clear. Secondly, it is uncertain how much of the deposited organic material is incorporated into benthic macrofaunal biomass or how important benthic food webs are to fisheries production. There is evidence which suggests that some benthic macrofaunal communities are food limited (Marsh and Tenore 1990) even in this relatively food-rich system and that predation is an important factor regulating benthic communities (Holland et al. 1980). Many potential interactions between benthic nutrient cycling and the eutrophication process remain unclear although there are tantalizing hints to the effect that strong positive and negative feedback mechanisms (i.e., destabilizing and stabilizing, respectively) dominate along the eutrophication gradient.

Explaining such phenomena will require a diversity of techniques. Experience in recent years has clearly indicated the value of generating long-term data sets collected at appropriate time scales. Such data sets are invaluable in developing understanding of linkages between system components and external factors believed to force the system. Due to climatic variability natural experiments are often contained within such data sets. In terms of developing clear linkages between benthic and pelagic system components and between external forcings such as nutrient loading and benthic components, it now appears that favoring more measurements at a limited number of locations provides more useful information than a strategy that emphasizes spatial at the expense of temporal coverage.

Secondly, utilization of improved technologies such as remote sensing and in-situ moorings designed to record selected environmental conditions would vastly improve our ability to determine the proper spatial and temporal scales, respectively, for study. It is possible to infer some cause-effect linkages from in-situ field studies and this has been the approach most frequently taken in studies in the Bay. Ecosystem simulation modeling, done in conjunction with such field studies, has proven to be a useful tool for integrating our understanding of community dynamics. Furthermore, there has been considerable success in using a variety of mesocosms as tools for conducting experimental ecosystem studies and this approach is capable of efficiently yielding many more insights on mechanisms controlling ecosystem processes. Finally, sufficient data are now available in both the published and gray literature to support intensive and reasonably detailed comparative studies of benthic-pelagic couplings among widely varying coastal ecosystems.

## *Management Implications*

The natural two-layer circulation of Chesapeake Bay waters creates conditions which favor the retention of nutrients in particulate forms, allowing this plant foodstuff to be efficiently used and reused before being buried, lost to gaseous forms or exported to the sea. The resulting high production of phytoplankton is a major factor contributing to anoxia during the warmer months each year.

This two-layer water flow in the Bay also separates upper and lower layers of the water column, effectively reducing the ability of oxygen to be transported from the atmosphere to the bottom waters and leading to depletion of oxygen (i.e., anoxia). Thus, this two-layer circulation, which is common in many other estuaries around the world, but particularly effective in Chesapeake Bay, makes the Bay very susceptible to the effects of nutrient enrichment on bottom water anoxia.

The processes by which particulate organic foods are deposited from the water column to the sediments are regulated by the nature of the planktonic food-chains, which are, in turn, controlled by the extent of nutrient enrichment in Chesapeake Bay. Any increases or decreases in nutrient inputs to the Bay are likely to significantly alter this process and result in shifts between the relative importance of fisheries production by demersal species (e.g., rockfish, weakfish, blue crabs) versus plankton based species (menhaden, bluefish). Any increases in particulate organic matter deposition will also contribute to greater depletion of oxygen from bottom waters.

In comparison with other coastal ecosystems, the relative production of benthic animals per unit algal production appears to be markedly inhibited in many regions of the Bay by the seasonal anoxia. Presumably, this process has affected also the production of key demersal fisheries, which are presently below their potential. Conversely, indirect evidence suggests that the exploitation of important suspension-feeding benthic animals (especially the oyster) may have contributed to the accumulation of phytoplankton biomass and the resulting reduction of Bay water clarity. This hypothetical relation needs further analysis before being translated into management strategies.

While nitrogen enrichment of Bay waters has contributed over recent decades to its eutrophication, this effect is mitigated by the natural bacterial processes of nitrification and denitrification, which transform nitrogen from useable forms (ammonium, nitrate) to unusable forms (dinitrogen gas) for estuarine phytoplankton. The coupled process of nitrification-denitrification is presently inhibited by hypoxic conditions during summer in many regions of the Bay. This amounts to a "positive feedback loop" in the Bay, where nitrogen enrichment leads to increased algal production, which leads to more anoxia which leads to less denitrification which leads to more nitrogen recycling, which leads to more algal production, which leads to more anoxia, and so on. Considering this relation in the inverse, even small reductions in nitrogen inputs to the Bay might lead to relatively large reversals of the eutrophication process by restoring nitrification-denitrification.

Strong evidence indicates that there is a residual effect of nutrient inputs to the Bay which operates on time scales of months to one or two years. In this case, physical processes of particle retention and relatively slow (compared to the water column) benthic processes of nutrient recycling combine to create a lag between nutrient inputs and some of the resulting primary production which is supported by those inputs. It appears that relatively high nutrient inputs during a high riverflow year support relatively high rates of primary production and benthic oxygen consumption in the following year. Most of these nutrients are not, however, retained in the estuary, such that ecological processes will respond rapidly to reductions in nutrient inputs. The question of whether a small residual effect can influence Bay ecology and for how long is central to predicting the impact of reducing nutrient loading to the estuary.

All of these mechanisms by which processes of benthic-pelagic coupling may affect the water quality and living resources of the Bay need to be taken into account in the development of models of water quality and ecosystem processes to predict the outcome of management strategies for improving environmental conditions in Chesapeake Bay.

## Acknowledgments

Much of this work reported here has been funded by the NOAA Sea Grant Program and by the Maryland Department of the Environment Monitoring Program ("Ecosystem Processes" Component). Additional support for developing this synthesis was provided by NSF LMER (BSR-8814272). Most of the results discussed here have been produced within the last three years, and many have yet to be published in the open literature. Therefore, numerous key figures and tables are presented in the present report to provide the reader access to the basic information. We gratefully acknowledge F. Holland and K. Sellner for allowing us to use their unpublished data. We also thank R. Bartleson and T. Randall for assistance with graphics.

## References

Andersen, V. and P. Nival. 1988. A pelagic ecosystem model simulating production and sedimentation of biogenic particles: role of salps and copepods. Mar. Ecol. Progr. Ser. 44:37-50.

Berner, R. A. 1980. Early Diagenesis. Princeton University Press, Princeton, New Jersey.

Billen, G. 1978. A budget of nitrogen recycling in North Sea sediments off the Belgian coast. Estuar. Coast. Mar Sci. 7: 127-146.

Billen, G. and C. Lancelot. 1988. Modeling benthic nitrogen cycling in temperate coastal ecosystems, pp. 341-378. In: J. Sørensen and T.H. Blackburn (eds.) Nitrogen Cycling in Marine Environments. John Wiley (SCOPE), New York.

Blackburn, T.H. and K. Henriksen. 1983. Nitrogen cycling in different types of sediments from Danish waters. Limnol. Oceanogr. 28:477-493.

Bloesch J. and R. D. Evans. 1982. Lead$^{210}$ dating of sediments compared with accumulation rates estimated by natural markers and measured with sediment traps. Hydrobiologia 92:579-586.

Bloesch, J. and N. M. Burns. 1980. A critical review of sedimentation trap techniques. Schweiz. Z. Hydrol. 42:15-56.

Blomqvist, S. and L. Hakanson. 1981. A review on sediment traps in aquatic environments. Arch. Hydrobiol. 91:101-132.

Bodungen, B.v., K.v. Brockel, V. Smetacek and B. Zeitzschel. 1981. Growth and sedimentation of the phytoplankton spring bloom in the Bornholm Sea (Baltic Sea). Kieler Meresforsch. Sonderh. 5:49-60.

Boynton, W. R., W. M. Kemp and C.G. Osborne. 1980. Nutrient fluxes across the sediment-water interface in the turbid zone of a coastal plain estuary, pp. 93-109. In: V.S. Kennedy (ed.) Estuarine perspectives. Academic Press, New York.

Boynton, W.R., W.M. Kemp, C.G. Osborne, K.R. Kaumeyer and M.C. Jenkins.1981. Influence of water circulation rate on in situ measurements of benthic community respiration. Mar. Biol. 65:185-190.

Boynton, W.R., W.M. Kemp and C.W. Keefe. 1982. A comparative analysis of nutrients and other factors influencing estuarine phytoplankton production. pp. 69-90. In V.S. Kennedy (ed.) Estuarine Comparisons. Academic Press, New York.

Boynton, W.R. and W.M. Kemp. 1985. Nutrient regeneration and oxygen consumption by sediments along an estuarine salinity gradient. Mar. Ecol. Progr. Ser. 23: 45-55.

Boynton, W.R., J. Garber, J. Barnes and W.M. Kemp. 1988. Ecosystem Processes Component. Level I Report to State of Maryland, Department of the Environment. Ref. No. (UMCEES) CBL 88-2. University of Maryland Center for Environmental and Estuarine Studies, Solomons.

Butman, C.A. 1986. Sediment trap biases in turbulent flows: Results from a laboratory flume study. J. Mar. Res. 44: 645-693.

Butman, C.A., W. D. Grant and K. D. Stolzenbach. 1986. Predictions of sediment trap biases in turbulent flows: A theoretical analysis based on observations from the literature. J. Mar. Res. 44:601-644.

Callender, E. and D. E. Hammond. 1982. Nutrient exchange across the sediment-water interface in the Potomac River estuary. Estuar. Coastal Shelf Sci. 15:395-413.

Carpenter, S. R., M. M. Elser and J. J. Elser. 1986. Chlorophyll production, degradation and sedimentation: Implications for paleolimnology. Limnol. Oceanogr. 31:112-124.

Caumette, P., M. Pagano and L. Saint-Jean. 1983. Répartition verticale du phytoplancton, des bactérie et du zooplancton dans un milieu stratifiyé en Baie de Biétri (Lagune Ebriyé, Côte d'Ivoire). Relations trophiques. Hydrobiologia 106:135-148.

Cederwall, H. and R. Elmgren. 1980. Biomass increase of benthic macrofauna demonstrates eutrophication of the Baltic Sea. Ophelia Suppl. 1:287-304.

Christensen, H. and E. Kanneworff. 1985. Sedimenting phytoplankton as major food sources for suspension and deposit feeders in the Øresund. Ophelia 24:223-244.

Christensen, J. P., W. M. Smethie and A. H. Devol. 1987. Benthic nutrient regeneration and denitrification on the Washington continental shelf. Deep-Sea Res. 34:1027-1047.

Christensen, P. B., L.P. Nielsen, N. P. Revsbech and J. Sørensen. 1989. Microzonation of denitrification activity in stream sediments as studied with combined oxygen and nitrous oxide microsensor. Appl. Env. Microbiol. 55:1234-1241.

Cloern, J E. 1982. Does the benthos control phytoplankton biomass in South San Francisco Bay? Mar. Ecol. Progr. Ser. 9:191-202.

Cohen, R. R. H., P. V. Dresler, E. J. Phillips and R. L. Cory. 1984. The effect of the Asiatic clam, *Corbicula fluminea*, on phytoplankton of the Potomac River, Maryland. Limnol. Oceanogr. 29:170-180.

Davies, J.M. 1975. Energy flow through the benthos in a Scottish sea loch. Mar. Biol. 31:353-362.

Day, J.W., C. Hall, W.M. Kemp and A. Yañez-Arancibia. 1989. Estuarine ecology. Wiley, New York.

D'Elia, C.F., J.G. Sanders and W.R. Boynton. 1986. Nutrient enrichment studies in a coastal plain estuary: Phytoplankton growth in large-scale, continuous cultures. Can J. Fish. Aquat. Sci. 43:397-406.

Doering, P.H., J.R. Kelly, C.A. Oviatt and T. Sowers. 1987. Effect of the hard clam *Merceneria mercenia* on benthic fluxes of inorganic nutrients and gases. Mar. Biol. 94:377-383.

Ducklow, H.W., S.M. Hill and W.D. Gardner. 1985. Bacterial growth and decomposition of particulate organic carbon collected in sediment traps. Continent. Shelf Res. 4:445-464.

Dyrssen, D. 1986. Chemical processes in benthic flux chambers and anoxic basin waters. Netherl. J. Sea Res. 20:225-228.

Edmondson, W.T. 1969. Eutrophication in North America, pp. 124-149. In: Eutrophication: Causes, consequences, correctives. Natl. Acad. Sci. Publ. 1700.

Falkowski, P.G., T.S. Hopkins and J.J. Walsh. 1980. An analysis of factors affecting oxygen depletion in the New York Bight. J. Mar. Res. 38:476-506.

Fisher, T.R., P.R. Carlson and R.T. Barber. 1982. Sediment nutrient regeneration in three North Carolina estuaries. Estuar. Coastal Mar. Sci. 14:101-116.

Flint, R.W. 1985. Coastal ecosystem dynamics: Relevance of benthic processes. Mar. Chem. 16:351-367.

Forsskahl, M., A. Laakkonen and J.-M. Leppanen. 1982. Seasonal cycle of production and sedimentation of organic matter at the entrance to the Gulf of Finland. Netherl. J. Sea Res. 16:290-299.

Frechette, M., C. A. Butman et al. 1989. The importance of boundary-layer flows in supplying phytoplankton to the benthic suspension feeder, *Mytilus edulis.* L. Limnol. Oceanogr. 34:19-36.

Garber, J.H. 1988. Benthic-pelagic coupling in the Chesapeake Bay, pp. 11-34. In: M.P. Lynch and E.C. Krome (eds.) Perspectives on the Chesapeake Bay: Advances in estuarine science. Publication No. 127, Chesapeake Bay Research Consortium, Gloucester Point, Virginia.

Gardner, W.D. 1979. Sediment trap dynamics and calibration: a laboratory evaluation. J. Mar. Res. 38:17-39.

Gardner, W.D. 1979. Field assessment of sediment traps. J. Mar. Res. 44: 41-52.

Gardner, W.S., B.J. Eadie, J.F. Chandler, C.C. Parrish and J.M. Malczyk. 1989. Mass flux and "nutritional composition" of settling epilimnetic particles in Lake Michigan. Can. J. Fish. Aquat. Sci. 46:1118-1124.

Gasith, A. 1975. Tripton sedimentation in eutrophic lakes—Simple correction for the resuspended matter. Verh. Internat. Verein. Limnol. 19:116-122.

Gerritsen, J. 1988. Biological control of water quality in estuaries; Removal of particulate matter by filter feeders. Proc. Oceans 88.

Gerritsen, J.J., A. Ranasinghe and A.F. Holland. 1988. Comparison of three strategies to improve water quality in the Maryland portion of Chesapeake Bay. Unpublished Report, Versar Inc., Columbia, Maryland.

Graf, G., W. Bengtsson, U. Diesner, R. Schulz and H. Theede. 1982. Benthic response to sedimentation of a spring phytoplankton bloom: Process and budget. Mar. Biol. 67:201-208.

Graf, G., R. Schulz, R. Peinert and L.-A. Meyer-Reil. 1983. Benthic response to sedimentation events during autumn to spring at a shallow-water station in the western Kiel Bight. Analysis of processes on a community level. Mar. Biol. 77: 235-246.

Grassle, J.F., J.P. Grassle, L.S. Brown-Leger, R.F. Petrecca and N.J. Copley. 1985. Subtidal macrobenthos of Narragansett Bay. Field and mesocosm studies of the effects of organic input on benthic populations, pp. 421-434. In: J. S. Gray and M.E.Christiansen (eds.) Marine Biology of Polar Regions and Effects of Stress on Marine Organisms. John Wiley, New York.

Grebmeier, J.M., C.P. McRoy and H.M. Feder. 1988. Pelagic-benthic coupling on the shelf of the northern Bering and Chukchi Seas. I. Food supply source and benthic biomass. Mar. Ecol. Progr. Ser. 48:57-67.

Hammond, D.G., C. Fuller, D. Harmon, B. Hartman, M. Korosec, L.G. Miller, R. Rea, S. Warren, W. Berelson and S.G. Hager. 1985. Benthic fluxes in San Francisco Bay. Hydrobiologia 129:69-90.

Hargrave, B.T. 1973. Coupling carbon flow through some pelagic and benthic communities. J. Fish. Res. Board Can. 1317-1326.

Hargrave, B.T. 1984. Sinking of particulate matter from the surface water of the ocean, pp. 155-178. In: J. E. Hobbie and P. J. LeB. Williams (eds.) Heterotrophic activity in the sea. Plenum Press, New York.

Hargrave, B.T. and G.A. Phillips. 1986. Dynamics of the benthic food web in St. Georges Bay, southern Gulf of St. Lawrence. Mar. Ecol. Progr. Ser. 31:277-294.

Harrison, W.G. 1980. Nutrient regeneration and primary production in the sea, pp. 430-466. In: P.G. Falkowski (ed.) Primary Productivity in the Sea. Plenum Press, New York.

Henriksen, K., J. Hansen, and T.H. Blackburn. 1981. Rates of nitrification, distribution of nitrifying activity and nitrate fluxes in different types of sediment from Danish waters. Mar. Biol. 61:299-304.

Henriksen, K. and W.M. Kemp. 1988. Nitrification in estuarine and coastal marine sediments: Methods, patterns and regulating factors, pp.207-249. In: T.H. Blackburn and J. Sørensen (eds.) Nitrogen Cycling in Coastal Marine Environments. John Wiley, New York.

Holland, A.F., N.K. Mountford and J.A. Mihursky. 1977. Temporal variations in Upper Bay mesohaline benthic communities: I. The 9-m mud habitat. Chesapeake Sce. 18:370-378.

Holland, A.F., N.K. Mountford, M.H. Hiegel, K.R. Kaumeyer and J.A. Mihursky. 1980. The influence of predation on infauna abundance in upper Chesapeake Bay. Mar. Biol. 57:221-235.

Holland, A.F. 1988. Long-term benthic monitoring for the Maryland portion of Chesapeake Bay. Rep. to Md OEP, Baltimore. Versar Inc., Columbia, Maryland.

Honjo, S. and M.R. Roman. 1978. Marine copepod fecal pellets: Production, preservation and sedimentation. J. Mar. Res. 36:45-57.

Hopkinson, C.S. 1987. Nutrient regeneration in shallow-water sediments of the estuarine plume region of the nearshore Georgia Bight, USA. Mar. Biol. 94:127-142.

Hopkinson, C.S. and R.L. Wetzel. 1982. In situ measurements of nutrient and oxygen fluxes in a coastal marine benthic community. Mar. Ecol Progr. Ser. 10:29-35.

Hopkinson, C.S., B. Sherr and W.J. Wiebe. 1989. Size fractionated metabolism of coastal microbial plankton. Mar. Ecol. Progr. Ser. 51:155-166.

Hutchinson, G.E. 1973. Eutrophication. Amer. Sci. 61:269-279.

Indrebø, G., B. Pengerud and I. Dundas. 1979. Microbial activities in a permanently stratified estuary. II. Microbial activities at oxic-anoxic interface. Mar. Biol. 51:305-309.

Jenkins, M.C. and W.M. Kemp. 1984. The coupling of nitrification and denitrification in two estuarine sediments. Limnol. Oceanogr. 29:609-619.

Jordan, S. 1985. Production and decomposition of fecal and pseudofecal material by the American oyster, *Crassostrea virginica.* PhD Thesis. University of Maryland, College Park.

Jørgensen, B.B. 1980. Season oxygen depletion in the bottom waters of a Danish fjord and its effect on the benthic community. Oikos. 34:68-76.

Jørgensen, B.B. 1983. Processes at the sediment-water interface, pp. 477-509. In: B. Bolin and R. B. Cook (eds.) The major biogeochemical cycles and their interactions. SCOPE publication, Paris.

Josefson, A.B. 1987. Large-scale patterns of dynamics in subtidal macrozoobenthic assemblages in the Skagerrak: Effects of a production-related factor? Mar. Ecol. Progr. Ser. 38:13-23.

Jumars, P.A. and K. Fauchald. 1977. Between community contrasts in successful polychaete feeding strategies, pp. 1-20. In: B. C. Coull (ed.) Ecology of Marine Benthos. Univ. of S. Carolina Press, Columbia.

Kelly, J.R. and S.W. Nixon. 1984. Experimental studies of the effect of organic deposition on the metabolism of a coastal marine bottom community. Mar. Ecol. Progr. Ser. 17:157-169.

Kemp, W.M. and W.R. Boynton. 1980. Influence of biological and physical factors on dissolved oxygen dynamics in an estuarine system: Implications for measurement of community metabolism. Estuar. Coast. Mar. Sci. 11:407-431.

Kemp, W.M. and W.R. Boynton. 1981. External and internal factors regulating metabolic rates of an estuarine benthic community. Oecologia 51:19-27.

Kemp, W.M. and W.R. Boynton. 1984. Spatial and temporal coupling of nutrient inputs to estuarine primary production: The role of particulate transport and decompostition. Bull. Mar. Sci. 35: 522-535.

Kemp, W.M., P.A. Sampou, J. Caffrey, M. Mayer, K. Henriksen and W.

R. Boynton. 1990. Ammonium recycling versus denitrification in Chesapeake Bay sediments. Limnol. Oceanogr. 35:1545-1563.

Kemp, W.M. and W.T. Randall. 1988. Ecosystem simulation modeling of processes relating nutrient cycling to oxygen depletion in Chesapeake Bay, pp. 29-35. In: D. O'Connor and J. Fitzpartick (eds.) Water Column State-Variables and Aquatic Processes. Hydroqual Publ., Mahwah, New Jersey.

Kennedy, V.S. and L. Breisch. 1981. Review of literature pertaining to the American oyster, *Crassostrea virginica*, in Chesapeake Bay. Maryland Sea Grant College Program, College Park.

Klump, J.V. and C.S. Martens. 1983. Benthic nitrogen regeneration, pp. 411-456. In: E. J. Carpenter and D. G. Capone (eds.) Nitrogen in the Marine Environment. Academic Press, New York.

Knauer, G.A. , D.M. Karl, J.H. Martin and C.N. Hunter. 1984. *In situ* effects of selected preservatives on total carbon, nitrogen and metals collected in sediment traps. J. Mar. Res. 42: 445-462.

Kuo, A.Y. and B.J. Neilson. 1987. Hypoxia and salinity in Virginia estuaries. Estuaries. 10:277-283.

Lehman, J.T. 1988. Hypolimnetic metabolism in Lake Washington: Relative effects of nutrient load and food web structure on lake productivity. Limnol Oceanogr. 33 (6 part 1):1334-1347.

Malone, T.C., W.M. Kemp, H.W. Ducklow, W.R. Boynton, J.H.Tuttle, and R.B. Jonas. 1986. Lateral variations in the production and fate of phytoplankton in a partially stratified estuary. Mar. Ecol. Progr. Ser. 32: 149-160.

Malone, T.C., L.H. Crocker, S.E. Pike and B.W. Wendler. 1988. Influences of river flow on the dynamics of phytoplankton production in a partially stratified estuary. Mar. Ecol. Progr. Ser. 48:235-249.

Marsh, A.G. and K.R. Tenore. 1990. The role of nutrition in regulating the population dynamics of opportunistic, surface deposit feeders in a mesohaline community. Limnol. Oceanogr. 35:710-724.

May, E.B. 1973. Extensive oxygen depletion in Mobile Bay, Alabama. Limnol. Oceanogr. 18:353-366.

Newell, R.I.E. 1988. Anoxia and sea nettles in Chesapeake Bay: Are they the result of overharvesting the American oyster, pp. 536-546. In: Understanding the Estuary: Advances in Chesapeake Bay Research, Conf. Proc., Publ. 129 (CBP/TRS 24/88). Chesapeake Bay Research Consortium Solomons, Maryland.

Nixon, S.W. 1981. Remineralization and nutrient cycling in coastal marine ecosystems, pp. 111-138. In: B.J. Neilson and L.E. Cronin (eds). Nutrients and Estuaries. Humana Press, New York.

Nixon, S.W. 1988. Physical energy inputs and the comparative ecology of lake and marine and marine ecosystems. Limnol. Oceanogr. 33(4 part 2):1005-1025.

Nixon, S.W. and M.E.Q. Pilson. 1983. Nitrogen in estuarine and coastal marine ecosystems, pp. 565-647. In: E.J. Carpenter and D.G. Capone (eds.) Nitrogen in the Marine Environment. Academic Press, New York.

Officer, C.B., T.J. Smayda and R. Mann. 1982. Benthic filter feeding: A natural eutrophication control. Mar. Ecol. Progr. Ser. 9:203-210.

Officer, C.G., R.B. Biggs, J.L. Taft, L.E. Cronin, M.A. Tyler and W.R. Boynton. 1984. Chesapeake Bay anoxia: Origin, development and significance. Science 23:22-27.

Oviatt, C.A. and S.W. Nixon. 1975. Sediment resuspension and deposition in Narragansett Bay. Estuar. Coastal Mar. Sci. 3:201-217.

Oviatt, C.A., A.A. Keller, P.A. Sampou and L.L. Beatty. 1986. Patterns of productivity during eutrophication: A mesocosm experiment. Mar. Ecol. Progr. Ser. 28:69-80.

Parsons, T.R., M. Takahashi and B. Hargrave. 1979. Biological oceanographic processes, pp. 203-207. Pergamon Press, Oxford.

Peterson, W.T. and H.G. Dam. 1988. The influence of copepod "swimmers" on pigment fluxes in brine-filled vs. seawater-filled particle traps (unpublished ms).

Richards, F.A. 1965. Anoxic basins and fjords, pp. 611-645. In: J.P. Riley and G. Skirrow (eds.) Chemical Oceanography. Academic Press, New York.

Rowe, G.T. 1971. Benthic biomass and surface productivity, pp. 441-453. In: J. Costlow (ed.) Fertility of the Sea, Volume 2. Gordon and Breach, New York.

Rowe, G.T. 1985. Benthic production and processes off Baja California, Northwest Africa and Peru: A classification of benthic subsystems in upwelling ecosystems. Int. Symp. Upw. W. Afr., Inst. Inv. Pesq. Barcelona, vol. 2: 589-612.

Rudnick, D.T. and C.A. Oviatt. 1986. Seasonal lags between organic carbon deposition and mineralization in marine sediments. J. Mar. Res. 44:815-837.

Schindler, D.W. 1978. Factors regulating phytoplankton production and standing crop in the wrold's fresh waters. Limnol. Oceanogr. 23:478-486.

Schindler, D.W. 1977. Evolution of phosphorus limitation in lakes. Science 195:260-262.

Seitzinger, S.P. 1988. Denitrification in freshwater and coastal marine ecosystems: Ecological and geochemical significance. Limnol. Oceanogr. 33(4, part 2):702-724.

Seliger, H.H., J.A. Boggs, W.H. Biggley. 1985. Catastrophic anoxia in the Chesapeake Bay in 1984. Science 228:70-73.

Seliger, H.H. and J.A Boggs. 1988. Long term patterns of anoxia in the Chesapeake Bay, pp. 570-583. In: Understanding the Estuary: Advances in Chesapeake Bay Research, Conference Proceedings. Publication No. 129 (CBP/TRS 24/88), Chesapeake Bay Research Consortium, Solomons, Maryland.

Smetacek, V. 1984. The supply of food to the benthos, pp. 517-547. In: M. J. Fasham (ed.) Flows of Energy and Materials in Marine Ecosystems. Theory and Practice. Plenum Press, New York.

Smetacek, V. 1985. Role of sinking in diatom life-history cycles: Ecological, evolutionary and geological significance. Mar. Biol. 84:239-251.

Smith, S.V., W.J. Wiebe, J.T. Hollibaugh, S.J. Dollar, S.W. Hager, B.E. Cole, G.W. Tribble and P.A. Wheeler. 1987. Stoichiometry of C, N, P and Si fluxes in temperate-climate embayment. J. Mar. Res. 45:427-455.

Sørensen, J. 1978. Denitrification rates in a marine sediment as measured by the acetylene inhibition technique. Appl. Environ. Microbiol. 36:139-143.

Steele, J.H. and I.E. Baird. 1972. Sedimentation of organic matter in a Scottish sea loch. Mem. Ist. Ital. Idrobiol. 29 (suppl):73-88.

Suess, E. 1980. Particulate organic carbon flux in the oceans: Surface productivity and oxygen utilization. Nature. 288:260-263.

Taft, J.L., W.R. Taylor, D.O. Hartwig and R. Loftus. 1980. Seasonal oxygen depletion in Chesapeake Bay. Estuaries 3:242-247.

Taguchi, S. 1982. Sedimentation of newly produced particulate organic matter in a subtropical inlet, Kaneohe Bay, Hawaii. Estuar. Coastal Shelf Sci. 14:533-544.

Townsend, D.W. and L.M. Cammen. 1988. Potential importance of the timing of spring plankton blooms to benthic-pelagic coupling and recruitment of juvenile demersal fishes. Biol. Oceanogr. 5:215-229.

Turner, R.E., W.W. Schroeder and W.J. Wiseman. 1987. The role of stratification in the deoxygenation of Mobile Bay and adjacent shelf bottom waters. Estuaries. 10:13-19.

Twilley, R.R. and W.M. Kemp. 1986. The relation of denitrification potentials to selected physical and chemical factors in sediments of Chesapeake Bay, pp. 279-293. In: D. Wolfe (ed.) Estuarine Variability. Academic Press, New York.

Ward, L.G. 1985. The influence of wind waves and tidal currents on sediment resuspension in middle Chesapeake Bay. Geo-Mar. Lett. 5:71-75.

Wassmann, P. 1984. Sedimentation and benthic mineralization of organic detritus in a Norwegian fjord. Mar. Biol. 83:83-94.

Williams, P.J. LeBl. 1981. Microbial contribution to overall marine plankton metabolism. Oceanol. Acta. 4:359-364.

# Conclusion

It is clear from the four chapters presented in this volume that our knowledge of the processes responsible for the creation and maintenance of the deep-water low dissolved oxygen phenomenon in Chesapeake Bay has greatly improved over the past few years. Yet as new scientific insights into the dynamics of the Bay emerge, new questions also arise. For example, because the introduction of fresh water into the Bay during the spring freshet is responsible for nutrient delivery as well as for setting up the stratified conditions in the Bay, clarifying the relative importance of each factor remains a difficult endeavor.

We now are aware that while oxygen consumption occurs fundamentally through biochemical mechanisms, the cumulative effect of these processes tends to be modulated by the phenomenon of physical stratification. Analysis of available data tends to support the conceptual framework that stratification inhibits oxygen supply to the lower-layer waters of the Bay. However, as has been pointed out in this book, this first-order model is subject to modifications that result from nonlinear, biochemical controls. In fact, biochemical controls are probably of key importance for establishing trends in the magnitude of the Bay's oxygen depletion in both time and space scales.

Further research into the details of small scale (spatial and temporal) physical phenomena and processes, and the enhancement of our ability to integrate these quantitative descriptions into an interacting, synergistic system, needs to be undertaken. Evidence has also been presented that interannual variation in summer anoxia is the result of climatic factors governing precipitation in the watershed, consequential fresh water inflow, nutrient loading and circulation. A clearer understanding of how changes, over time, in the patterns of watershed use (such as deforestation) have influenced runoff, erosion (sediment transport) and nutrient inputs to the Bay system will provide a better assessment of the relative impacts of natural versus anthropogenic events on ecosystem function.

Our understanding of the interaction between nutrients and biota in the Bay has also increased significantly. Observed patterns of nutrient distribution and recycling suggest that annual phytoplankton production within the Bay is nitrogen-limited. On smaller temporal and spatial scales phosphorus or silicon can be limiting, especially in spring, when nitrogen input concentrations are high. Recycled nutrients support the major portion of this annual production.

Observation of seasonal variations in phytoplankton nutrient status, as well as phytoplankton and zooplankton abundance, imply a seasonal partitioning of phytoplankton production among microbial and metazoan food webs. Microbial abundance and activity in the Bay are extremely high during the periods of peak phytoplankton biomass and productivity, and this "microbial loop" cycles much of the Bay's organic material at these times. Yet the mechanisms that govern this partitioning of phytoplankton production among microbial and metazoan food webs remain unclear.

One of the most important results of the research described in this volume has been the realization that the heterotrophic microbial component of the Bay's ecosystem plays a major role in regulating water quality and oxygen dynamics. Bacterially mediated decomposition of organic material in the water column is sufficient at most times to account for observed oxygen demands. However, the specific sources, and concentrations, of bacterially labile organic matter in the Bay are not totally known. In addition, it is not clear whether the relationships which have been identified in the Bay proper also exist in the major tributaries.

Clearly, there is a need for long timeseries data sets. Analysis of these data sets will provide the first indications of whether the Bay is responding to present and future management strategies. Just as importantly, these data will help provide insight into the distinction between natural processes and anthropogenic alterations in the environment. Additionally, the use of other techniques such as paleoecology, remote sensing, mesocosm studies, and stable isotope analysis hold promise for providing further important information necessary for the understanding (and management) of Chesapeake Bay.

From a management perspective, one of the most important conclusions of this analysis is that nutrient inputs, primarily from nonpoint sources delivered to the Bay by the Susquehanna River, support an accumulation of phytoplankton biomass that is in excess of the aerobic, oxygen assimilation capacity of the Bay under current conditions of climate and watershed use. That is, most of the Bay appears to be "nutrient saturated" or "organically saturated."

This conclusion implies that nutrient loading reductions may not be reflected in improved water quality (such as increased deep water oxygen) until nutrient inputs are reduced to some unknown level. This "level" is probably not constant because of interannual changes in climatic factors, such as precipitation, mixing events or storms, and patterns of watershed use, as well as "feedback" factors such as enhanced denitrification, which could result from improved dissolved oxygen conditions. Secondly, because of the Bay's variability, improvements in water quality may not be readily discernible for several years following the implementation of a particular management practice. (This despite the finding that the Bay, as a whole, does not have a long "nutrient memory," and that annual inputs are generally utilized within a year.) Some of the highest priorities for new research include better understanding of feedback and time lags.

Because of this potential "lag" in separating a real trend from inherent system "noise," cost-effective management practices must either proceed slowly or must be increasingly dependent on the application of empirical and theoretical models. Empirical observation of ecosystem function and the development of sound ecosystem theory are thus critical to the development of ecologically and economically sound management schemes.

Nutrient management decisions intended to target the mesohaline reach of the Bay should be based on criteria that reflect the relationships between nutrient input (not concentration), phytoplankton biomass (not productivity), and the temporal and spatial extent of oxygen depletion. It should also be recognized that accumulations of phytoplankton biomass which fuel oxygen depletion in this region of the Bay appear to be more responsive to nitrogen loading (and possibly silica inputs) than to phosphorus loading. Increased reductions in phosphorus inputs due to nutrient control strategies may, however, move the system towards phosphorus limitation.

Finally, nutrient enrichment appears to affect the nature of the planktonic food-chain within the water column, that is, the relative importance of microbial versus metazoan pathways. These trophic pathways in turn dictate the mechanisms by which particulate organic material is deposited to the sediment. Consequently, alteration in nutrient inputs can result in a shift in the food chains which support higher trophic levels (with implications for fishery production), and in the relative importance of demersal or bottom feeding fish and shellfish versus plankton-based species. Increased organic input to the benthos, in the absence of benthic consumers, may contribute to greater depletion of oxygen from the bottom waters. Ultimately, such inputs will adversely impact the fisheries dependent on such benthic habitats.

# Glossary

**Algae.** A group of non-vascular aquatic plants, most of which have chlorophyll; often referred to interchangeably as phytoplankton.

**Allochthonous.** Referring to particulate material produced outside a particular ecosystem (e.g., oak leaves falling into a Bay tributary) and imported into the system; as opposed to autochthonous material.

**Anoxia.** The absence of dissolved oxygen in the aquatic environment.

**Anthropogenic.** Referring to the influence of human beings on the environment.

**Autocthonous.** Referring to particulate material produced within a designated ecosystem (e.g., phytoplankton production) as opposed to allocthonous material.

**Autotrophic processes.** Referring to biochemical processes which fix the sun's energy, resulting in manufactured organic compounds from inorganic substances (e.g., photosynthesis).

**Bacterivores.** Organisms which feed primarily on bacteria (e.g., many members of the protozoa).

**Bacterioplankton.** Free living bacteria in the water column.

**Benthic organisms.** The organisms living in or on the bottom of aquatic environments; includes bacteria, seagrasses, worms, oysters. Referred to collectively as members of the benthos.

**Benthic-pelagic coupling.** Referring to the functionary interconnections and pathways through which energy and organic material are transferred between the pelagic ecosystem and the benthic ecosystem.

**Biomass.** The mass of living matter within a given area per volume of the environment; often referred to interchangeably as standing crop.

**Biochemical oxygen demand (BOD).** A measure of the oxygen required by aerobic organisms and chemical reactions within a particular environment, or environmental sample.

**Carnivores.** Consumers of primarily animal matter.

**Chlorophyll *a*.** A group of green pigments found in plant cells that are active in harnessing energy during photosynthesis; often used as a measure or index of phytoplankton biomass.

**Chop-Pax transect.** An east-west sampling transect across the Chesapeake Bay extending from the Choptank to the Patuxent rivers.

**Diatom.** Major phytoplankton group enclosed within secreted silicon frustules, or shells; representatives of the taxa Bacillariophyceae.

**Dinoflagellates.** Major phytoplankton group possessing flagella, or whip-like swimming-structures; representatives of the taxa Dinophyceae.

**Ecology.** The science of the interaction of organisms with their environment.

**Ecosystem.** The system created by the interaction of a community of organisms with their environment.

**Estuary.** A semi-enclosed body of water which has free connection with the open sea and within which sea water is measurably diluted by freshwater.

**Euphotic zone.** The vertical region of the water column through which sufficient sunlight can penetrate for photosynthesis to occur.

**Eutrophic.** Refers to aquatic environments which are characterized as relatively productive, that is, they exhibit high concentrations of nutrients; as opposed to oligotrophic.

**Eutrophication.** The process involving overenrichment with nutrients in aquatic ecosystems: initially may involve increased productivity but with continued enrichment will eventually result in deteriorating environmental quality.

**Food chain.** Refers to the flow of energy linearly through an ecosystem from ecological level to ecological level (i.e., phytoplankton to zooplankton).

**Food web.** Refers to the flow of energy through an ecosystem via successive steps or trophic levels (i.e., organism A to organism B); often this flow of energy is not strictly linear.

**Freshet.** The large spring pulse of fresh water entering the Chesapeake Bay generally resulting from spring rains and snow melt.

**Halocline.** The vertical gradient in water column salt concentration commonly found in estuaries; a physical manifestation of the typical two-layer circulation found in Chesapeake Bay.

**Herbivores.** Consumers of primarily phytoplankton or other plant material.

**Heterotrophic processes.** Referring to the processes through which organic matter is broken down, thereby releasing energy for growth and metabolism.

**Hypoxia.** The presence of very low dissolved oxygen in the aquatic environment; usually considered to be dissolved oxygen concentrations ranging between 2 to 0 $mgO_2\ L^{-1}$.

**Labile organic matter.** Organic material which is relatively easy to breakdown via heterotrophic processes; as opposed to refractory organic matter.

**Macrobenthos.** Benthic organisms generally larger than one millimeter.

**Macroplankton.** Plankton ranging in size between 20 and 200 micrometers.

**Meiobenthos.** Benthic organisms ranging in size between 100 to 1000 micrometers.

**Mesohaline.** An estuarine environment having a salt content between 5 and 20 parts per thousand.

**Microbenthos.** Benthic organisms ranging in size between 1 and 100 micrometers.

**Microplankton.** Plankton ranging in size between 20 and 200 micrometers.

**Nannoplankton.** Plankton ranging in size between 2 and 20 micrometers.

**Nutrients.** Inorganic chemical species required for growth; major nutrients necessary for phytoplankton growth include nitrogen, phosphorus, carbon and silicon.

**Nutrient loading.** The amount of a particular nutrient which is introduced into an ecosystem over a given time period ($kg\ yr^{-1}$).

**Oligohaline.** An estuarine environment having a salt content between 0.5 and 5 parts per thousand.

**Oligotrophic.** Refers to aquatic environments which are characterized as relatively unproductive, that is, impoverished in nutrients; as opposed to eutrophic.

**Pelagic organisms.** The organisms living within the water column of aquatic environments; includes bacteria, phytoplankton, zooplankton, fish.

**Photosynthesis.** The autotrophic process where light energy is used to derive energy rich organic compounds from inorganic building blocks.

**Polyhaline.** An estuarine environment having a salt content between 20 and 35 parts per thousand.

**Primary producers.** Organisms such as phytoplankton and seagrasses which are autotrophic.

**Primary production.** The amount of organic matter produced via the primary producers; often referred to interchangeably as plant biomass.

**Primary productivity.** Primary production (i.e., plant biomass) produced within a given volume, or under a given area, per unit time (e.g., gC $m^{-3}d^{-1}$ or mgC $m^{-2}d^{-1}$).

**Pycnocline.** The region of the water column characterized by a large vertical gradient in density; commonly found in estuaries, it is a manifestation of the typical two-layer circulation found in Chesapeake Bay.

**Refractory or organic matter.** Referring to organic matter which is relatively resistant to biochemical breakdown; as opposed to labile organic material.

**Salinity.** The weight in grams of dissolved salts in one kilogram of estuarine/sea water; usually expressed as parts per thousand.

**Turbidity.** A term used to describe the degree of transparency or opaqueness produced by suspended particulate matter in the water column.

**Two-layer circulation.** The circulation pattern characteristic of many estuaries where freshwater flows toward the sea and over the denser seawater flowing up the estuary. Mixing occurs between these layers throughout the estuary.

**Zooplankton.** Diverse group of minute animals that freely drift in the water,

# Appendix

## Low Dissolved Oxygen Research Projects

The following list includes only Sea Grant projects supported by pass-through funds orginating from a Chesapeake Bay appropriation to the National Oceanic and Atmospheric Administration. Additional projects supported by core Maryland and Virginia Sea Grant funds are not listed. We include only those projects which focus on the mechanisms responsible for, and the maintenance of, the hypoxic/anoxic bottom water phenomenon in Chesapeake Bay. Other research projects dealing with the effect of the low dissolved oxygen water in ecologically and economically important species have also been funded but are not listed here.

*Title:* Chesapeake Bay Dissolved Oxygen Dynamics: Roles of Phytoplankton and Microheterotrophs
*Duration:* 1985-87
*Principal Investigators:* Robert B. Jonas and Thomas C. Malone

*Title:* Relative Roles of Benthic vs. Pelagic Oxygen-Consuming Processes in Establishing and Maintaining Anoxia in Chesapeake Bay
*Duration:* 1985-86
*Principal Investigators:* W. Michael Kemp and Walter R. Boynton

*Title:* The Role of Heterotrophic Bacteria in Generating and Maintaining Anoxia in Chesapeake Bay
*Duration:* 1985-86
*Principal Investigator:* Hugh W. Ducklow

*Title:* Implications of Microzooplankton Grazing on Carbon Flux and Anoxia in Chesapeake Bay
*Duration:* 1985-86
*Principal Investigators:* Lawrence W. Harding, Jr. and Kevin G. Sellner

*Title:* Physical and Biological Processes Regulating Anoxia in Chesapeake Bay: Zooplankton Distributions, Grazing and Respiration
*Duration:* 1985-86
*Principal Investigator:* Michael R. Roman

*Title:* Contribution of Sulfur Cycling to Anoxia in Chesapeake Bay
*Duration:* 1985-86
*Principal Investigator:* Jon H. Tuttle

*Title:* Sources of Biochemical Oxygen Demand in the Chesapeake Bay: The Role of Macrophyte Detritus and Decomposition Processes
*Duration:* 1985-86
*Principal Investigators:* Joseph C. Zieman, Stephen A. Macko and Aaron L. Mills

*Title:* Hypoxia in Virginia's Saline Waters: A Review and Analysis of Historic Data
*Duration:* 1985
*Principal Investigator:* Lawrence W. Haas

*Title:* Bacterial Production and Population Dynamics in the Mesohaline Reach of Chesapeake Bay
*Duration:* 1987-88
*Principal Investigator:* Hugh W. Ducklow

*Title:* Grazing and Production of Zooplankton in Chesapeake Bay
*Duration:* 1987-88
*Principal Investigator:* Michael R. Roman

*Title:* Role of Grazing by Nanozooplankton and Microzooplankton on Bacteria and Phytoplankton in Chesapeake Bay
*Duration:* 1987-88
*Principal Investigator:* Evelyn J. Lessard

*Title:* Integration of Ecological Processes Relating Nutrient Enrichment to Anoxia and Trophic Dynamics in Chesapeake Bay: A Simulation Study
*Duration:* 1987-88
*Principal Investigator:* W. Michael Kemp

*Title:* Chronic Effects of Seasonally Low Dissolved Oxygen on the Resource Value of the Bay Bottom
*Duration:* 1987-89
*Principal Investigators:* Robert J. Diaz, Linda Schaffer and Mark Luckenbach

*Title:* Predation Impact of Jellyfish and Ctenophores on Plankton Populations in Chesapeake Bay
*Duration:* 1988
*Principal Investigator:* Jennifer E. Purcell

*Title:* The Role of Dissolved Organic Constituents from Anaerobic Decomposition in Oxygen Dynamics in the Water Column
*Duration:* 1988-89
*Principal Investigators:* Linda K. Blum and Aaron L. Mills

*Title:* Passive Artificial Ventilation of Hypoxic Estuarine Benthic Environments
*Duration:* 1988-89
*Principal Investigators:* L. Don Wright, John D. Boon, J. P.-Y. Maa, D. M. Bushnell and G. C. Greene

*Title:* The Impact of Oyster Abundance on Water Quality in Sub-Estuaries of Chesapeake Bay
*Duration:* 1989
*Principal Investigators:* Robert B. Jonas, Jon H. Tuttle and Steven Coon

*Title:* The Production and Fate of Phytoplankton Biomass in the Mesohaline Reach of the Chesapeake Bay
*Duration:* 1989
*Principal Investigators:* Daniel J. Conley, Thomas C. Malone and Patricia M. Glibert

*Title:* Trophic Dynamics in Chespeake Bay: Ecological and Biochemical Approaches to the Study of Flows of Carbon from Phytoplankton to Bacterioplankton
*Duration:* 1989
*Principal Investigators:* Hugh W. Ducklow, Madilyn Fletcher and Peter C. Griffith

*Title:* Plankton Trophodynanics in Chesapeake Bay: Bacterivory and Herbivory by Nano- and Microzooplankton
*Duration:* 1989
*Principal Investigator:* George B. McManus

*Title:* Plankton Trophic Dynamics in Chesapeake Bay: Models and Budgets of Carbon-Nutrient Cycling
*Duration:* 1989
*Principal Investigator:* W. Michael Kemp

*Title:* Ship-Time Support and Coordination
*Duration:* 1985-86, 88
*Principal Investigator:* Thomas Malone